TesselMania!™ Math Connection

Lois Edwards

Kevin Lee

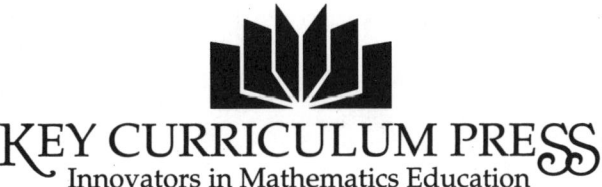

TesselMania! Math Connection

Authors: Lois Edwards and Kevin Lee

Editors: Greer Lleuad and Kristin Sallee

Production: Lois Edwards and Luis Shein

Special thanks to: the educators who helped test these materials, including Jim Kearns in Andover, Massachusetts; the teachers attending the summer course at The Geometry Center in Minneapolis, Minnesota; and those participating in workshops at Key Curriculum Press in Berkeley, California. We are particularly appreciative of the comprehensive review provided by Alice Foster of Omaha, Nebraska. Thanks also to the secondary students enrolled in the summer mathematics course at Macalester College in St. Paul, Minnesota, for testing the activities and providing valuable feedback.

Butterfly design on cover by Charolyn Kapplinger © 1994, created with *TesselMania!* ™ © 1994 MECC. All rights reserved. Used with permission of MECC.

TesselMania! is a trademark of MECC.

10 9 8 7 6 5 4 3   98 97 96

© 1995 by Key Curriculum Press. All rights reserved.
ISBN 1-55953-076-6

Key Curriculum Press
P.O. Box 2304
Berkeley, California 94702
510-548-2304
editorial@keypress.com

Limited Reproduction Permission
The publisher grants the teacher who purchases *TesselMania! Math Connection* the right to reproduce the material for use in his or her own classroom. Unauthorized copying of *TesselMania! Math Connection* constitutes copyright infringement and is a violation of federal law.

Printed in the United States of America.

Contents

About This Activity Book 1
   Objectives 1
   Connections to NCTM *Standards* 1
   Materials 2
   Student Prerequisites 2
   The Mathematics 2

Links to Your Curriculum 4
   Sequence 4
   Computers and Hands-on Activities 5
   In-Class and Project Activities 6
   Related Materials 6

Background Information 8
   M. C. Escher 8
   Heinrich Heesch 8

Student Activities and Teacher's Notes 11
   1. Tessellation Exploration 11
   2. Classifying Tiles 17
   3. Creating Tile Sides 25
   4. Translation and Rotation Tilings 29
   5. Glide-Reflection Tilings 35
   6. Combination Tilings 39
   7. Triangle and Hexagon Tilings 43
   8. Escher's Classification System 49
   9. Reverse Engineering 55
   10. Minimal Coloring 63
   11. Coordinate Connections 69

Transparency Masters 87
   Transparency Master: Activity 1 87
   Transparency Master: Activity 4 89
   Transparency Master: Activities 5 and 6 91
   Transparency Master: Activity 7 93
   Transparency Master: Activity 8 95
   Transparency Master: Activity 9a 97
   Transparency Master: Activity 9b 99
   Transparency Master: Activity 11 100

About This Activity Book

Objectives

The eleven activities in this book enable high school students to investigate the mathematics behind tessellations, like those created by artist M. C. Escher, using the *TesselMania!* software. Rather than emphasizing the connection between art and mathematics that is intrinsic to M. C. Escher's tessellations, these activities focus on the underlying transformational geometry.

Students explore and apply three types of transformations: translation, rotation, and glide reflection. The activities combine structured mathematical investigations and computer software animation of the transformations. This approach develops students' intuitive understanding and strengthens their ability to analyze and apply transformations.

Students examine two different tessellation classification systems. One system was developed by mathematician Heinrich Heesch. The other was created by M. C. Escher himself to help him create and classify his drawings. In studying these systems, students strengthen, apply, and extend their knowledge of transformations, as well as experience the power of classification systems.

Additionally, they apply problem-solving skills, including working backwards, collecting data, organizing data in a table, and looking for patterns and relationships.

All of the activities stress communicating mathematics. Students are encouraged to represent their findings graphically and verbally. They will work in pairs and explain and defend their conclusions to their partners and to the class. Most of the investigations include short writing activities to summarize results.

Connections to NCTM *Standards*

These activities support the goal of the *Curriculum and Evaluation Standards for School Mathematics*, published by NCTM in 1989. The first four goals—problem solving, communication, reasoning, and mathematical connections—are inherent to most of the activities.

Activities 9 and 10 specifically involve problem solving. Activities 2 and 4–8 focus on classification systems and their notation as a means of communication. They also present connections between the Heesch and the Escher classification systems. In Activity 10, students make and test conjectures. Activity 11 relates tessellations to coordinate geometry. Tessellations and transformations in synthetic geometry, which are featured in the *Standards* (page 158), are the focus of all of these activities. Coordinate geometry, algebraic equations, and matrices are represented in Activity 11, which links computer algorithms to transformations. Activity 10 provides another discrete mathematics activity.

About This Activity Book

Materials

This book includes blackline masters for student activities and teacher's notes for each activity. The blackline masters may be reproduced for classroom use. The teacher's notes contain background information, suggested in-class demonstrations and discussions, and sample results. Transparency masters are also included.

Additional materials are sometimes required. These are specified in the teacher's notes pages, and include transparency film, transparency pens, and tracing paper or patty paper (translucent squares of paper used to separate hamburger patties—these can be obtained from restaurant supply distributors or from Key Curriculum Press).

These activities require the use of the *TesselMania!* software. They are more advanced mathematically than the lessons found in the manual that accompanies the software. You will want to refer to the manual for software information, for preliminary activities, and for references for further reading.

Student Prerequisites

Students need to have had hands-on experience with tiling the plane, using simple geometric shapes such as quadrilaterals and triangles. They need to have worked with the transformations of basic shapes, and they should understand and be able to apply terminology for transformations: *translation, rotation, reflection*. At a minimum, they should be comfortable with the informal terms: *slide, turn, flip*.

Activity 1 is a hands-on activity in which students create tessellations by tracing shapes. It is designed to review and extend students' previous work with tessellations. If your students have *not* had prior experience with this type of exploration, they should begin with the introductory lessons included in the manual that accompanies the *TesselMania!* software.

Some of the activities have specific prerequisites, which are listed in the teacher's notes pages. In particular, Activity 11 requires algebra and coordinate graphing experience. Note that Activity 4 is a prerequisite for Activities 5–11.

The Mathematics

All the activities in this book involve transformations. The three transformations that are used are translation, rotation, and glide reflection.

The sides of a basic tile, such as a square, are transformed to create a motif. This motif may or may not resemble a physical object, such as a butterfly or a lizard. The motif is then transformed to tile the plane.

Students investigate the types of transformations. They relate the transformations of the tile sides to the transformations used to create the tessellation. The Heesch System (see Background Information) provides a useful method for classifying the types of tessellations investigated. The system devised by M. C. Escher exemplifies a different approach to classification.

About This Activity Book

The *TesselMania!* software greatly facilitates these explorations. Through computer animation, it enables students to see the movement inherent in the transformations.

Some activities connect transformations and tessellations to other areas of mathematics. Activity 10 relates tessellations to map coloring. Activity 11 applies coordinate geometry concepts to the transformations of tessellations. This activity also relates transformations to computer science, which uses a coordinate grid and transformation equations to draw graphic shapes on a computer screen

Throughout these activities, students are involved in collecting and recording data from their computer investigations. They are looking for patterns and relationships, and they are summarizing and making conjectures based on their findings.

Since the activities in this book focus on the mathematics rather than the art of tessellations, the tiles are geometric shapes rather than recognizable objects (except in the activity that involves reverse engineering). This is intended to help students concentrate on the geometric concepts. It should not diminish the connection between art and mathematics that is inherent to the study of tessellations.

Links to Your Curriculum

1. Tessellation Exploration
2. Classifying Tiles
3. Creating Tile Sides
4. Translation and Rotation Tilings
5. Glide-Reflection Tilings
6. Combination Tilings
7. Triangle and Hexagon Tilings
8. Escher's Classification System
9. Reverse Engineering
10. Minimal Coloring
11. Coordinate Connections

Sequence

In general, these activities should follow the suggested sequence. Many of the later activities are dependent on the experience gained in completing the earlier activities.

In Activity 1: Tessellation Exploration, students create tessellations using tracing paper or patty paper. This is an important prerequisite for all the other activities. If you prefer, you can substitute a similar activity from your textbook or other materials. It should be hands-on, require students to record the types of moves they use to create the tessellation, and involve shapes more complicated than regular polygons.

Activity 2: Classifying Tiles is an introduction to the Heesch classification system. It is an optional paper and pencil activity. Activity 3: Creating Tile Sides provides practice using *TesselMania!* to create tiles of specific shapes. It is optional.

Activities 4–7 all focus on the Heesch classification system. They are prerequisites for the later activities. An understanding of the Heesch system is essential for the remaining activities. Students *must* complete all or most of these activities before going on to the others. You may want to do some in class and assign some for independent work. (See below.)

Activities 4: Translation and Rotation Tilings, 5: Glide-Reflection Tilings, and 6: Combination Tilings all focus on quadrilateral-based tiles. Activity 7: Triangle and Hexagon Tilings deals with triangle and hexagonal tiles.

After students have completed Activities 4–7, they may do any of the four other activities (8–11) in any order. Because of time constraints or student needs, you may choose to omit Activity 8: Escher's Classification System. This is a more challenging activity, but is valuable in comparing and contrasting the Escher system with the Heesch system.

Activity 9: Reverse Engineering is highly motivating. Students analyze the transformations used in a tessellation and try to re-create it. It is based on the Heesch system and does *not* require use of the Escher classification system.

Activity 10: Minimal Coloring appeals to most students. It involves data collection and analysis of the data. Students use the software to color tessellations of each of the Heesch types. They look for a pattern to determine which types require two colors and which require three colors. This is a valuable experience in conjecturing from specific examples.

Activity 11: Connecting Coordinate Geometry requires the most mathematical experience, including some background in algebra and coordinate geometry. Students write equations to determine coordinates for image points under transformations. You may choose to do only parts of this activity, depending on your students' background.

If time is very limited, Activities 1, 4, and 9 can comprise a short but effective unit.

Computers and Hands-on Activities

Computer software activities provide an intermediate step between manipulative activities and abstract, theoretical investigations. Software activities should be preceded by hands-on experimentation with tessellations. Some students may want to use tracing paper or patty paper and transparency film throughout these activities to complement their computer work.

Computers and the *TesselMania!* software must be available to students. A computer lab with a demonstration system and a computer for every two students is ideal. However, three or four computers in the classroom to be shared by groups of students is adequate. The teacher-led demonstrations require a computer system that includes a projection device. It is possible, but less convenient, to get by with just one computer system with a projection device. In this case, groups of students use it on their own as well as participating in demonstrations for the whole class.

Be sure to demonstrate how to operate *TesselMania!* before students use it for the first time. Follow the suggestions in the software manual under "Starting a New Tessellation" and create a simple tile using translation. Be sure to use the Show Me features. You do *not* need to point out the Heesch-type labels at this time.

Students need an opportunity to explore the *TesselMania!* software for 20–30 minutes before they begin Activities 4 or 3. Otherwise, they may find it difficult to focus on the structured investigation in the activity and may choose instead to experiment with the software features.

Links to Your Curriculum

In-Class and Project Activities

All of these activities can be done during class. Most activities begin with a demonstration. Most end with a discussion of your students' results. However, you may want to assign some activities or parts of activities as homework, or use some activities as either required or optional projects.

Activity 1 can be done individually after students create one traced tessellation together as an example. Additional tessellations can be assigned as homework. Activities 2 and 3 are best done in class.

Activity 4 should be done in class. The teacher-led demonstration is crucial. Activities 5, 6, and 7 can be done independently by pairs of students with access to computers and software. The first part of each problem can be done without a computer, but recording the tessellation moves requires *TesselMania!*

Activity 8 begins with an important demonstration. Students should at least start the activity in class.

For Activity 9, you can do the worked example as a whole class and then assign the other problems as independent work. This will require that students have access to *TesselMania!* If this activity is used as an independent project, be sure to give students a copy of the teacher's notes, which give directions and a worked example.

Activity 10 requires intensive use of *TesselMania!* However, students might collect their data in the computer lab and finish their summary and conjectures outside of class. Advanced students can do this activity on their own as a project.

Activity 11 will require some in-class introduction, but parts of it can be done as homework. You can omit the sections that deal with matrices if your students have not used matrices. You might choose to use the section on translation matrices as a lead-in to a matrix unit.

Related Materials

Visions of Symmetry

Visions of Symmetry: Notebooks, Periodic Drawings, and Related Work of M. C. Escher by Doris Schattschneider, published by W. H. Freeman and Co., is a rich source of information on Escher's tessellations and his classification system. It is available in many public libraries and can be purchased from Key Curriculum Press.

The Geometer's Sketchpad

You may want students to create tessellations using The Geometer's Sketchpad software, published by Key Curriculum Press. They can create motif tiles and use the transformation options to make tessellations.

Links to Your Curriculum

Discovering Geometry

Chapter 7 of this textbook, written by Michael Serra and published by Key Curriculum Press, covers transformations and tessellations. You may want to use this material before beginning the activities contained in *TesselMania! Math Connection*.

Patty Paper Geometry

Written by Michael Serra and published by Key Curriculum Press, this book of hands-on paper folding using patty paper includes activities on transformations and tessellations.

Other Texts

Many recently-published geometry textbooks include activities with geometric transformations—check your curriculum for the appropriate placement of these activities. You may want to supplement your current text with materials that provide your students with hands-on transformation activities.

Background Information

M. C. Escher

M. C. Escher (1889–1972) was a Dutch artist whose tessellation drawings are largely based on recognizable images. He imaginatively turned geometric shapes into such creatures as birds, lizards, and fish. Most of his tessellation art was created from about 1937 to 1958.

Although he was not a mathematician, Escher found that he needed to pursue the mathematics of transformations to help him create tessellation drawings. Over a period of three years, he developed a classification system for tessellations, based on the transformations needed to move the motif (original tile) to its adjacent positions. This helped him investigate the possible types of tessellations.

He identified ten principal "systems," or types, of tessellations and labeled them with Roman numerals. The first nine systems deal with quadrilaterals, and the tenth uses triangles. They are described in a chart that lists the translations, rotations, and glide reflections for each system (see *Visions of Symmetry*, page 62). Each tessellation is labeled with a Roman numeral to identify its system and with a superscript letter that indicates the type of quadrilateral used as the basic shape. For example, I^D represents a tessellation based on a square (D) with translations in both transversal and diagonal directions (System I).

Escher considered himself an artist, not a mathematician. In fact, he had been a poor mathematics student. However, his interest in tessellation art led him into the world of mathematics, and he made significant contributions in both areas.

For further information on M. C. Escher, see the sections "The Product in Detail" and "References" in the *TesselMania!* software manual. A highly recommended resource for teachers is the book *Visions of Symmetry: Notebooks, Periodic Drawings, and Related Work of M. C. Escher* by Doris Schattschneider, published in 1990 by W. H. Freeman and Co. Although it is written at a mathematically sophisticated level, students who are pursuing project work in tessellations will find parts of it accessible and valuable.

Heinrich Heesch

Heinrich Heesch was a German mathematician. In 1932, he investigated and classified shapes of tiles that could tessellate the plane. However, this work was not published until 1963 when Heesch and a colleague wrote a book on the topic. In June 1963, Heesch sent a letter to M. C. Escher. He expressed his admiration of Escher's work and included a chart of his own classification system with a description of his work. (See *Visions of Symmetry*, page 44, for further details on the correspondence between Escher and Heesch.)

Background Information

It is worth noting that Heesch created his system to be published and used by anyone interested in the mathematics of tessellations. Escher devised his system for his own use in creating works of art.

Heesch's system is based on how the transformations of the sides of an individual tile relate it to the tiles that surround it in a tessellation. He considered asymmetric, isohedral tiles. *Isohedral* means that for any two congruent tiles, there is a transformation (an isometry) that will move one tile exactly onto the other. He identified 28 types—*TesselMania!* includes 15 of these types. (The Heesch system is briefly described in the Heesch Types feature in the Options menu of the software.)

The Heesch type of a tessellation consists of a multiletter code. The number of letters represents the number of sides of the original tile shape. The letters are *T*, *G*, and *C* with subscripts: *T* stands for *translation*, *G* for *glide reflection*, and *C* for *center of rotation*. (Students who find it difficult to associate the letter *C* with the center of rotation might want to think of rotation as *circular* turning.)

The sequence of letters is obtained by "traveling" clockwise around the edges of a tile and assigning the appropriate letter to each side. In the simplest example, TTTT, each of the four sides is formed by translation.

Subscripts are used with G to indicate corresponding pairs of glide reflections. For example, $G_1 G_1 G_2 G_2$ is distinct from $G_1 G_2 G_1 G_2$ because in the first case the top side glide-reflects to the *adjacent* side, but in the second it glide-reflects to the *opposite* side. (See the animated examples in *TesselMania!*)

Rotations are indicated by C with a subscript that indicates the number of degrees of rotation. The letter C itself stands for 180°, C_3 stands for 120° (360° ÷ 3), and C_4 stands for 90° (360° ÷ 4). If the side is formed by C, the center of rotation is the midpoint of that side. Half of the side rotates onto the other half. For C_3 and C_4, the center of rotation is the vertex between that side and the adjacent side (moving clockwise). The side rotates onto the adjacent side. (See the animated examples in *TesselMania!*)

The Heesch types introduced in this book are listed below.

Triangle Tiles

    CCC   CGG

Quadrilateral Tiles

TTTT	$G_1 G_1 G_2 G_2$	CCCC
TCTC	$G_1 G_2 G_1 G_2$	$C_3 C_3 C_3 C_3$
GTGT	GGCC	$C_3 C_3 C_6 C_6$
	GCGC	$C_4 C_4 C_4 C_4$

Hexagonal Tiles

    TTTTTT $C_3 C_3 C_3 C_3 C_3 C_3$

Activity 1: Tessellation Exploration

Explore shapes that tile the plane and the moves used to make the tilings.

Overview

You already know that some regular polygons will tile the plane and that other polygons will not. What about more complicated shapes? Can they also tile the plane? How do you need to move the shapes to make the tilings?

The graphic artist M. C. Escher (1898–1972) was intrigued with the idea of creating a single shape (tile) that could be copied and then used to cover a surface without any gaps or overlaps. These coverings or tilings of the plane are also called *tessellations*.

In this activity, you will trace each tile several times to create a tessellation if that is possible. You will also record the moves you made.

Steps

- Copy the tile onto tracing paper or patty paper. (Use a different piece of paper for each tile. You can turn the paper over if you need to.)
- Move the paper to copy the tile again, fitting it against the first tile.
- Continue until you are convinced that the tile will or will not tessellate.
- Write the type of move *inside* each traced tile.
- Summarize your findings. Describe the types of moves you used.

The first tile has been started for you as an example (the tracing is reduced in size).

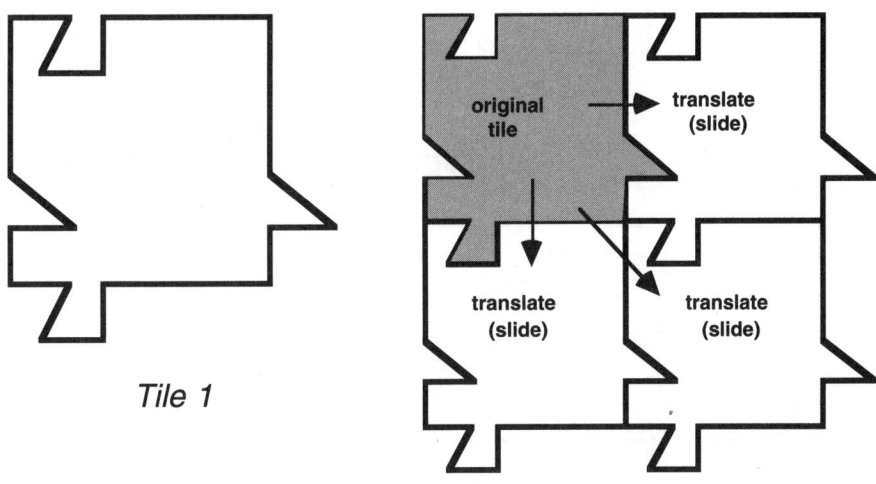

Tile 1

partial tracing of tessellation

TesselMania! Math Connection

Tessellation Exploration
page 2 of 2

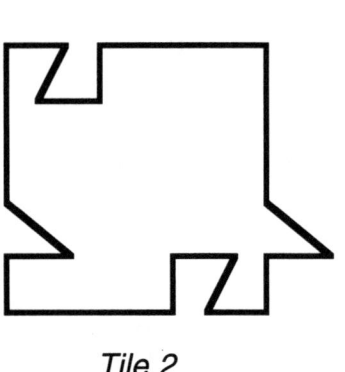
Tile 2

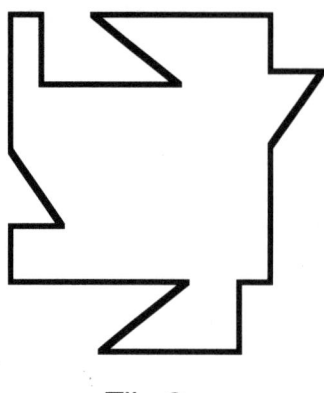
Tile 3

Tile 4

Results

	Tile 1	Tile 2	Tile 3	Tile 4
Moves				
Does it tessellate?				

Summary

What moves did you use? Do you see any patterns here? Describe them.

_____
_____
_____
_____
_____
_____
_____

Tessellation Exploration

Activity 1: Teacher's Notes

Objectives

To explore shapes that tile the plane and the moves used to make the tilings
To identify the transformations (moves) used in creating a tessellation
To differentiate among the moves: translation, rotation, and glide reflection
To begin to see relationships between the shape of a tile and its tessellation

Materials

Tracing paper or patty paper (four pieces per student or student pair—see page 2 for more on patty papers)
Student Activity Sheet 1 (one copy per student pair)
Transparency Master: Activity 1

Prerequisites

Students should know from experience that some regular polygons will tile the plane (quadrilaterals, triangles, hexagons) and that other polygons will not (regular pentagons). In this activity they explore more complicated shapes, like those they can create with *TesselMania!*

Students should be familiar with the terms *tiling the plane* and *transformation*. They should also have had informal experience with transformations, perhaps using the terms *slide, turn,* and *flip*. The term *tessellation* is introduced. Students should already know or learn the terms *translation, rotation,* and *glide reflection* during these activities. The classification systems are based on these mathematical terms for transformations.

Notes

Have students work in pairs or small groups for this activity. Each student can do half of the tracings and then present results to the partner(s). You may want to ask students to predict which tiles they think will tessellate before they experiment. However, some students may quickly see that Tile 2 does not tessellate, which may spoil the exploration for others. If you have advanced students who can "see" the results for Tiles 1 and 2, suggest that they start with Tiles 3 and 4.

In Class

Read or tell in your own words the Overview information on the activity sheet. Point out to students that you will do the first tile as an example and will follow the steps that are listed in the activity.

Using two transparencies on the overhead projector, demonstrate how to trace a shape and then move the tracing "paper" (the top transparency) to make the next tile. Ask students to describe the ways they could move the tracing paper. Be sure that they realize that sliding *is* a move (they may think that sliding is "nothing"). Be sure they mention turns (rotations) and flips (reflections) also. Write these three terms on the board or the overhead for emphasis.

Tessellation Exploration
Teacher's Notes • page 2 of 4

Suggest that students trace the first tile near the center of their tracing paper. Encourage students to trace tiles (nine in all) all around the first tile unless they see that the tile does not tessellate.

Note that if flips are needed, students will have to trace on both sides of the paper.

Be sure that students notice and record the transformations needed during the tracing. Have students write the name of the move inside each traced tile. Encourage more advanced students to specify the center of rotation and the axis of reflection.

Students may start to notice that there are relationships between the shape of the tile and the transformations needed to tessellate. They may see that each tile was created by applying transformations to a square. These relationships will be explored further in Activities 2, 3, and 4.

Use the Discussion Questions suggested in these notes to help students reflect on their work.

Results

Three of the four tiles do tessellate (Tiles 1, 3, and 4). Tile 2 does not tessellate. Even though the "peg" on the bottom of Tile 2 *does* fit into the top side, this creates overlapping tiles when the adjacent tile is placed on the left side.

Students will study Heesch types in later activities. For your information, Tile 1 is an example of Heesch type TTTT. Tile 3 is $G_1 G_2 G_1 G_2$. Tile 4 is TCTC.

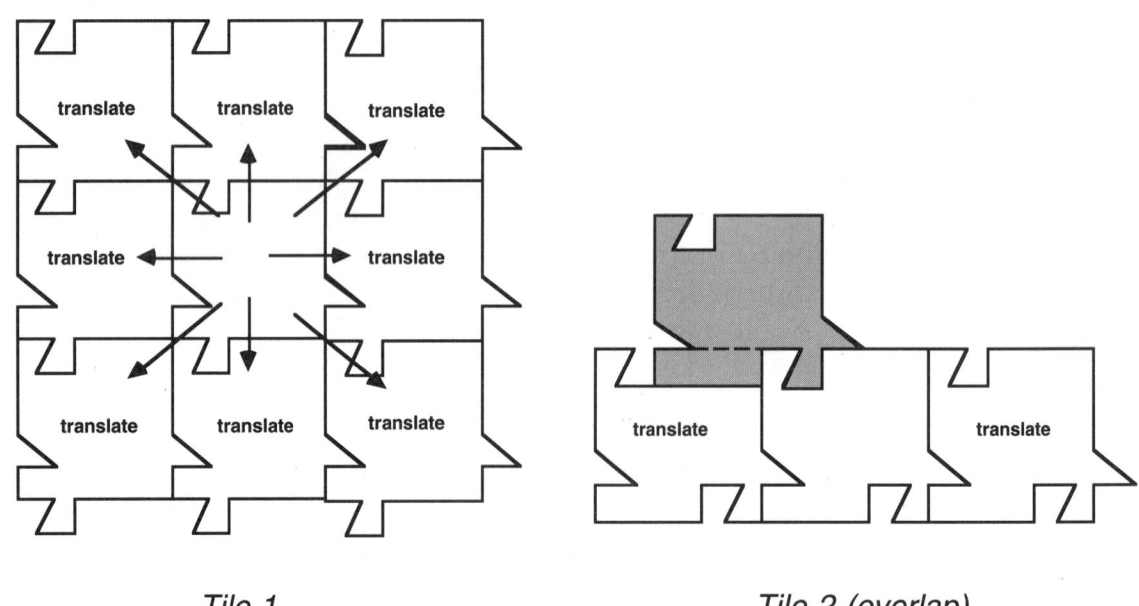

Tile 1   *Tile 2 (overlap)*

Tessellation Exploration
Teacher's Notes • *page 3 of 4*

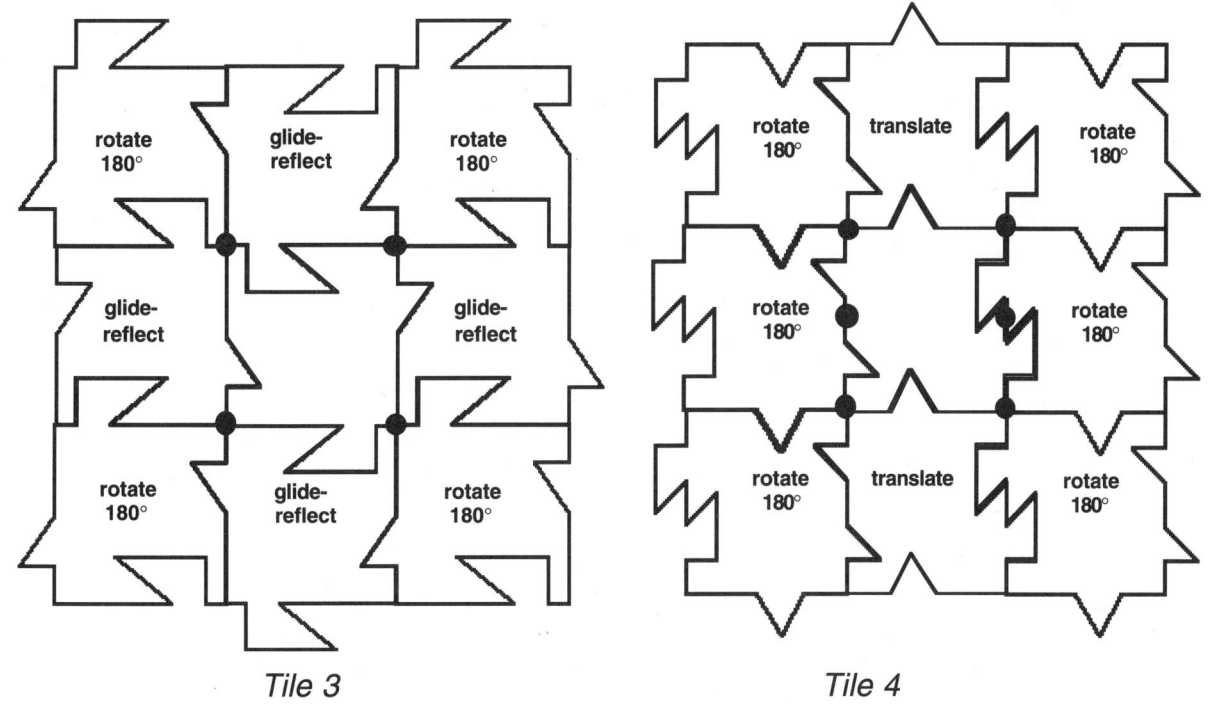

Tile 3   *Tile 4*

	Tile 1	Tile 2	Tile 3	Tile 4
Moves	slide up slide down slide left slide right slide up & left slide up & right slide down & left slide down & right	slide left slide right cannot fit on top or bottom	flip/slide flip/slide flip/slide flip/slide rotate (180°) rotate (180°) rotate (180°) rotate (180°)	slide up slide down rotate (180°) rotate (180°) rotate (180°) rotate (180°) rotate (180°) rotate (180°)
Does it tessellate?	yes	no	yes	yes

Discussion Questions

Have one or two student groups present their results. Ask if any groups had different results. If so, encourage students to present their evidence during the discussion.

How many tiles did you trace to be sure that a tile tessellated?

The traced tiles should encircle the first tile to show that the whole plane could be covered. Note that this is not *proof*, but it is convincing evidence that the tile tessellates.

Tessellation Exploration
Teacher's Notes • page 4 of 4

Show the transparency of Tile 1 tessellated. Write the moves on each adjacent tile as students report them from their experiment. Show that the eight adjacent tiles are sufficient.

What did you discover about Tile 2? Why didn't Tile 2 tessellate?

There is no way to move the paper so that one tile can fit above or below the first tile without overlapping other tiles. It *is* possible to make a horizontal string of tiles. Students may describe this tile as having the top and bottom cuts made *into* the shape, while the left cut is *in* and the right is *out*. If some students are not convinced, demonstrate the overlapping on the overhead.

What moves did you make with your tracing paper to do the tessellations?

The moves are slide, turn, and flip or translate, rotate, and reflect. Note that the order of the moves will differ depending on which position students chose to place the next copy of the tile. The order is not significant in this activity. Students will probably not differentiate between reflection and glide reflection. You can point out the glide reflections in Tile 3.

Did any of these operations involve more than one type of move?

Yes, Tile 3 requires a flip and a slide in both vertical and horizontal directions. The combination of a flip and a slide (reflection and translation) is a *glide reflection*. It is important that students understand glide reflection and that it involves two moves. Also, the translation must be along the line of reflection. You may want to point out that *glide* is a synonym for *slide* or *translation*.

Did any of these operations (moves) change the size or shape of the original tile?

No, all copies are congruent to the original. These operations are called *isometries* (*iso* means "equal") because they preserve the angle measure and side length. Mathematicians have classified four isometries of the plane: translation, rotation, reflection, and glide reflection.

Summary

What moves did you use? Do you see any patterns here? Describe them.

Tile 1 uses only translations; the top tile side translates to the bottom and the right side to the left. Tile 3 uses glide reflections and rotations (180°); the tile sides are formed by glide reflections. Tile 4 uses rotations (180°) and translations; the sides are formed by translations and rotations.

Activity 2: Classifying Tiles

Explore a system for classifying tiles that tessellate.

Background

Mathematician Heinrich Heesch became fascinated with the various types of tiles that will tessellate. He had seen M. C. Escher's tessellation art and decided to find a way to classify tiles. In this activity, you re-create his classification system and evaluate it.

Steps

- Examine each of the tiles below. Each was originally a square.

- Start at the top of each tile and move clockwise around the sides. Determine what kind of transformation created that side from the square. (Often a pair of sides is made using one transformation.)

- Label each side with a letter: T for translation, C for rotation, G for glide reflection. For rotations, use C for 180° and C_4 for 90°. If there are two different glide reflections, use G_1 and G_2. These labels are the symbols Heesch used. Mark the centers of rotations.

- Underneath the tile, write a four-letter code to represent the four transformations. This code is called the *Heesch type*. Tile 1 has been done as an example. (More than one tile may have the same Heesch type.)

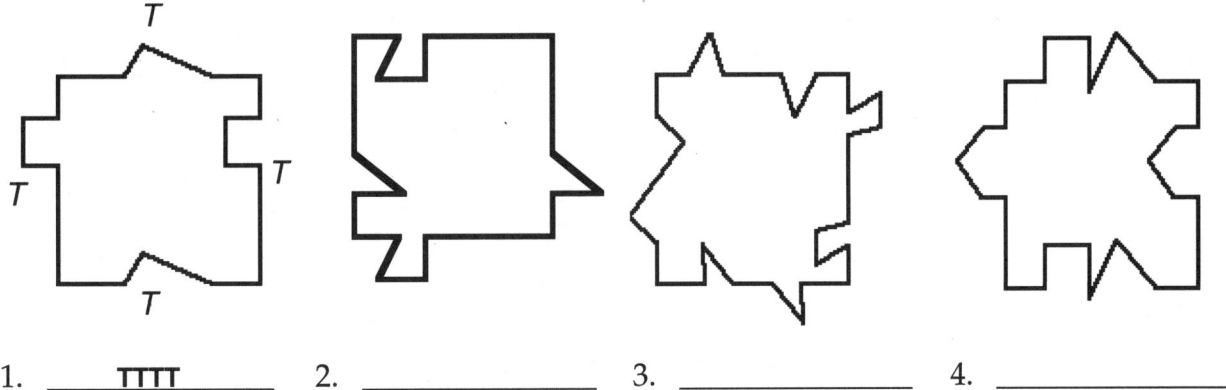

1. ___TTTT___ 2. _____ 3. _____ 4. _____

TesselMania! Math Connection

Classifying Tiles
page 2 of 3

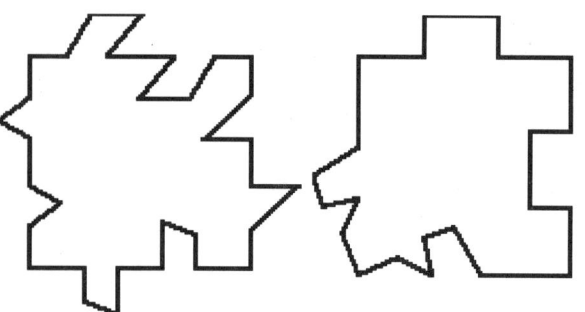 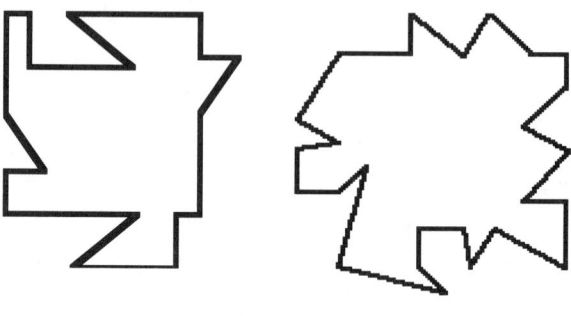

5. _____ 6. _____ 7. _____ 8. _____

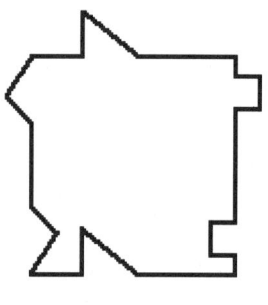 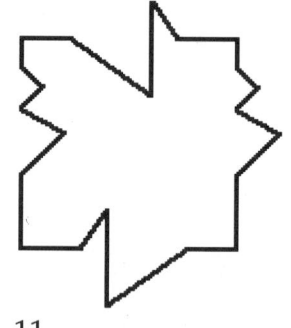

9. _____ 10. _____ 11. _____ 12. _____

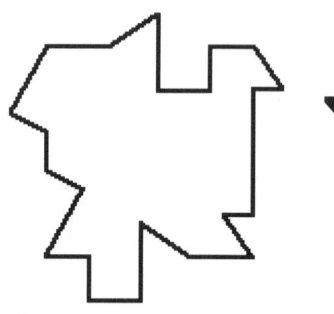  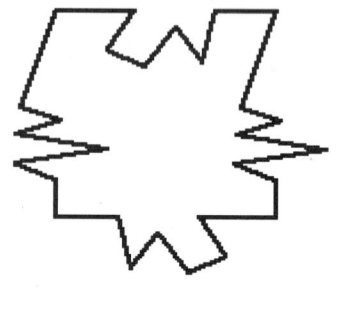

13. _____ 14. _____ 15. _____ 16. _____

Classifying Tiles

Summary

Imagine that your class partner was absent during this activity. Explain to him or her what the four-letter Heesch type means for square tiles. (Hint: One of the Heesch labels is explained for you. You explain the other labels.)

For square tiles . . .

If there is a T in the Heesch type, then the side translates to the opposite side.

Application

Since there is a Heesch type TCTC, could there also be a type that is TTCC? Explain.

Evaluation

List two things you find useful about the Heesch classification system. List two things you think could be improved.

Creative Thinking

How would you improve the Heesch system for classifying tiles?

Classifying Tiles
Activity 2: Teacher's Notes

Objectives

To explore a system for classifying tiles that tessellate
To investigate the Heesch classification system
To evaluate the Heesch system
To identify the Heesch type for a tile
To describe a tile, given its Heesch type

Materials

Student Activity Sheet 2

Prerequisites

Students should have had experience with the three types of transformations used to create the sides of tessellating tiles: translation, rotation (180° and 90°), and glide reflection.

It is suggested that students have had 20–30 minutes of exploration with *TesselMania!* before doing this activity. Seeing the transformation of sides by using the Show Me feature of the software will help students understand this activity.

Background

Mathematician Heinrich Heesch became fascinated with the various types of tiles that will tessellate. He had seen M. C. Escher's tessellation art and decided to find a way to classify tiles.

Escher himself created a different classification system, which is explored in Activity 8. However, most people find Heesch's system easier to understand and more helpful. The *TesselMania!* software uses the Heesch system.

Although students may find Heesch's symbols somewhat obscure at first, they will need to learn this system to investigate the mathematics behind tessellations. The Summary writing activity will demonstrate whether students understand the Heesch system. The Evaluation and Creative Thinking parts of this activity help students reflect on this system and vent any dissatisfaction they may have with it.

This activity deals only with transformations of *square* tiles. In other activities, students will also explore nonsquare tiles, but square tiles are sufficient for learning the Heesch system notation.

In Class

Present a brief introduction about Heinrich Heesch and the value of classifying all the many types of tiles that will tessellate. Read to the class the Background paragraph on the students' page.

Read each of the Steps to the students. On the board or overhead, work through a simple example of a TTTT tile, such as Tile 1 on the student sheet.

Classifying Tiles
Teacher's Notes • page 2 of 4

Be sure students understand that *each* side must have a letter, even though the top and bottom may be the result of the same translation, as in Tile 1. Emphasize that the letter Heesch used for rotation is C (not R). Point out that C without a subscript stands for a rotation of 180°. The code C_4 stands for 90°. (Ask students if they can explain this. A full rotation is 360°, and 90° = 360° ÷ 4.) Empathize with students if they object to Heesch's symbols, but ask them to play along and use them anyway. These symbols will prove useful in later activities.

Have students work on the activity sheets in pairs. Each pair should check their results with another pair before doing the writing on the last page. Don't be concerned about the subscripts used in Tiles 7 and 15; the reason they are needed is rather subtle and not essential to understanding the basic Heesch system. (There are additional symbols for rotations of 30°, 60°, and 120°, but these will be introduced later.)

Have students write their responses in sentence form. Encourage them to write so that someone who has not done the activity can understand them.

Results

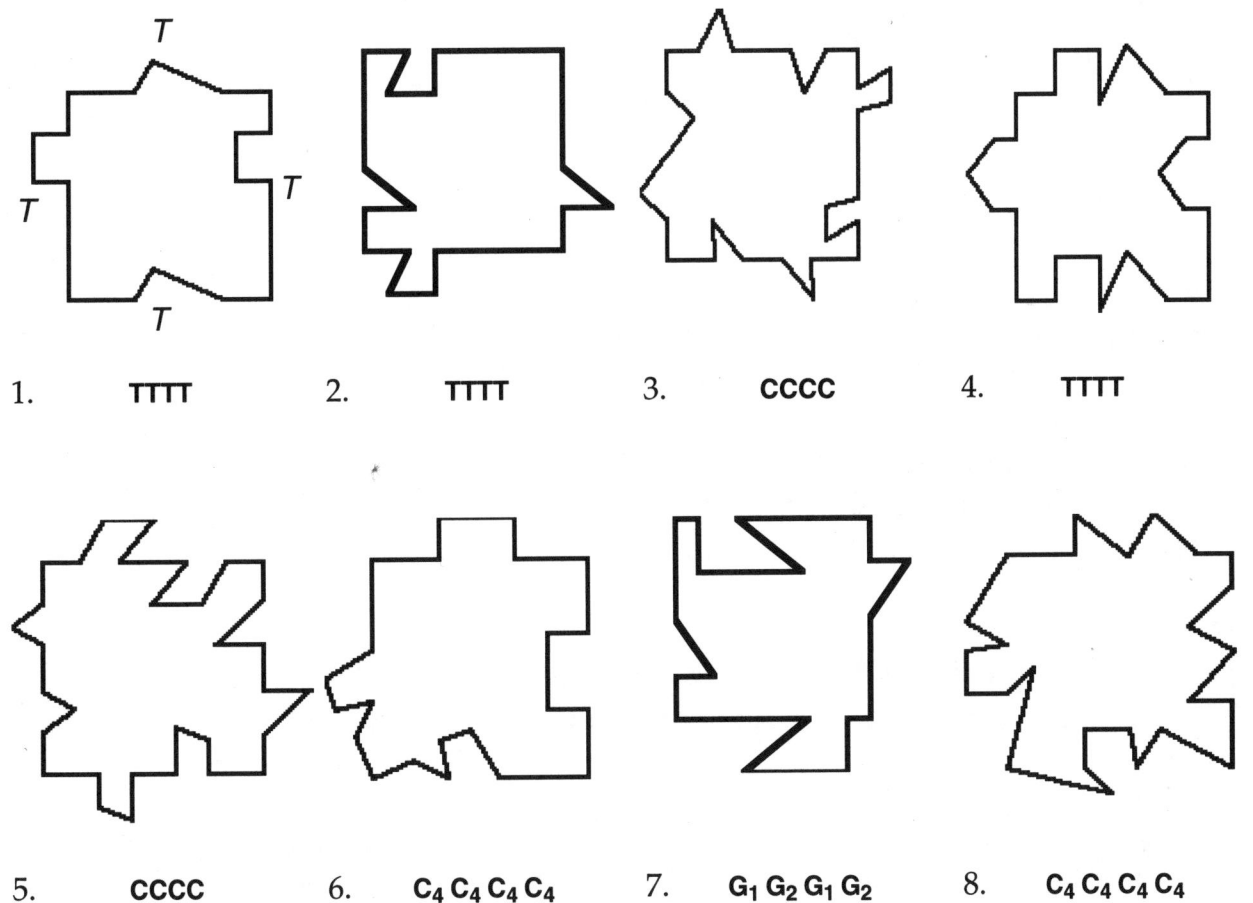

1. **TTTT** 2. **TTTT** 3. **CCCC** 4. **TTTT**

5. **CCCC** 6. **$C_4 C_4 C_4 C_4$** 7. **$G_1 G_2 G_1 G_2$** 8. **$C_4 C_4 C_4 C_4$**

Classifying Tiles
Teacher's Notes • page 3 of 4

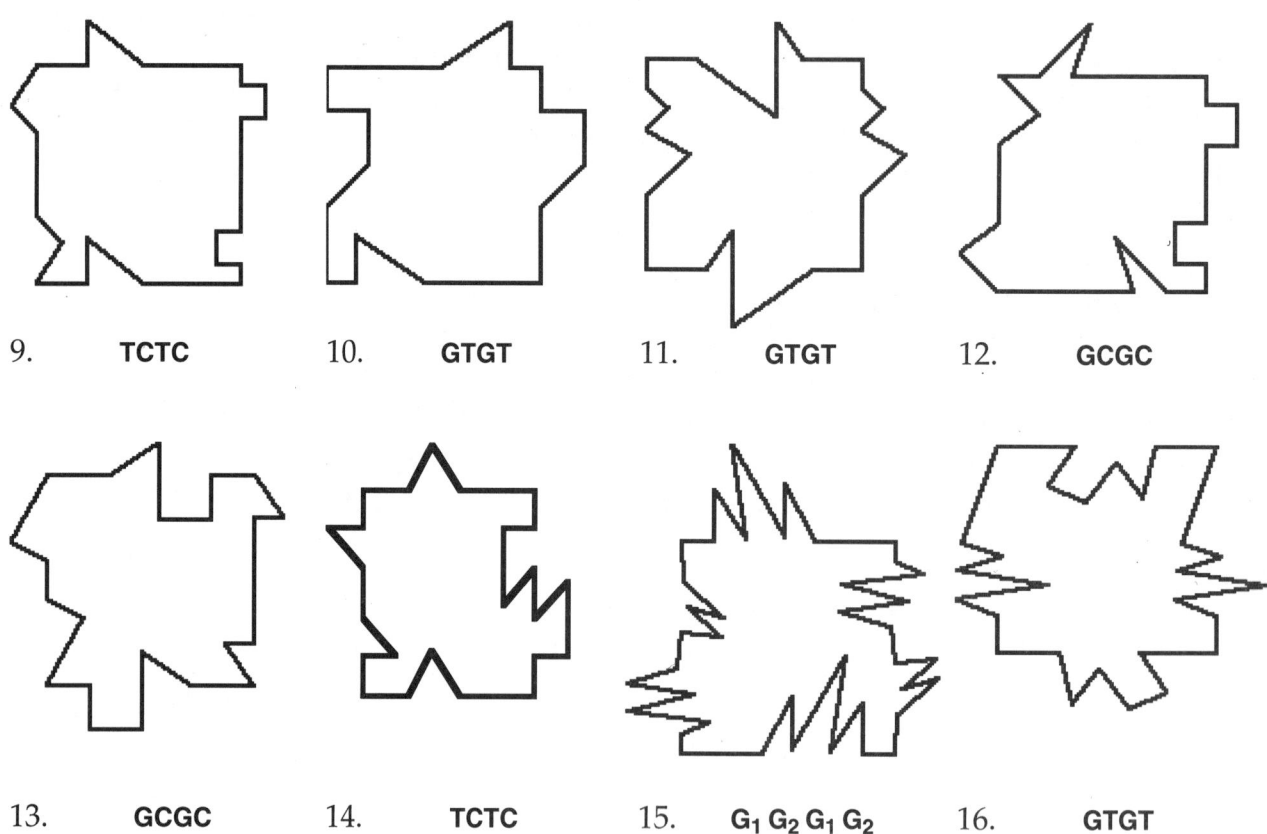

9. **TCTC** 10. **GTGT** 11. **GTGT** 12. **GCGC**

13. **GCGC** 14. **TCTC** 15. **$G_1 G_2 G_1 G_2$** 16. **GTGT**

Summary

Imagine that your class partner was absent during this activity. Explain to him or her what the four-letter Heesch type means for square tiles.

If there is a T in the Heesch type, then the side translates to the opposite side.

If there is a C, then half of that side rotates 180° around the side center.

If there is a C_4, then the side rotates 90° around the corner to the adjacent side.

If there is a G, then the side glide-reflects to the opposite side.

Application

Since there is a Heesch type TCTC, could there also be a type TTCC? Explain.

No, translation can move one side only to the opposite side, not the adjacent side.

Evaluation

List two things you find useful about the Heesch classification system. List two things you think could be improved.

Useful

There is one letter for each side, so you know the number of tile sides.

Classifying Tiles
Teacher's Notes • page 4 of 4

There is a different letter for each type of transformation.
The rotations are marked with subscripts to show the amount of rotation.

Needs Improvement

The letter C is used for rotation.

Subscripts, like 4, are used for the amount of rotation.

Some letters have subscripts but not all.

The code does not show which two sides transform into each other.

Creative Thinking

How would you improve this Heesch system for classifying tiles?

Use the letter R for rotation.

Use subscripts of 180 and 90 with R to show the amount of rotation.

Use subscripts to mark which two sides transform into each other ($T_1 T_2 T_1 T_2$).
Use an arrow to show the direction of the translation or glide reflection.

Extension

(This is not included on the student activity sheet.) Organize these Heesch types (or *all* of the Heesch types used in *TesselMania!*) in a way that makes sense to you.

Assessment

The Summary activity can be used for assessment. You might also ask students to write the Heesch type for the tiles in Activity 1.

Activity 3
Creating Tile Sides

Explore ways to create tiles using *TesselMania!*

Background

In *TesselMania!* you can change the original square tile by using the tack tool (this tool looks like a thumb tack) to add points to a side. Then you can drag those points to change the shape of the side. In this activity you will explore ways to arrange and move those points. You will investigate the relationship between the number of points on a side and the shape of that side.

Steps: Part 1

- Examine each of the tile sides.
- Decide how many points you will need to add to the tile to re-create this shape. Mark those points on the tile side. (Do *not* count the two endpoints of the side.)
- Start *TesselMania!* and choose the Translation option from the main menu.
- Re-create the tile side on the computer. Check to see that it matches the printed side.

number of tack points _____

number of tack points _____

number of tack points _____

number of tack points _____

number of tack points _____

number of tack points _____

number of tack points _____

number of tack points _____

Steps: Part 2

- Using the Translation option in *TesselMania!*, create as many different sides as you can using just one tack, then two tacks, and so on.
- Sketch three of your most interesting sides in the table on the next page. (You decide which are most interesting.)

Creating Tile Sides
page 2 of 2

Interesting Tiles Sides

1 Tack			
2 Tacks			
3 Tacks			
4 Tacks			
5 Tacks			

Summary

Compare your tile sides to those done by other students. Look for similarities and differences. Write a summary of what you noticed about the shape of a tile side and the number of tack points on it.

_____
_____
_____
_____

Challenge

How many tack points do you need to make a curved tile side? Make a conjecture and test it using *TesselMania!* Refine your conjecture as needed.

_____

Theoretically speaking, how many points are needed to make a curve? Is this answer the same as the one above? Explain why these two answers are the same or different.

_____
_____
_____

Creating Tiles Sides
Activity 3: Teacher's Notes

Objectives

To explore ways to create tiles using *TesselMania!*
To develop skills in using the tack tool in *TesselMania!*
To determine the number of tacks needed to create a given shape
To identify relationships between the number of tack points and the side's shape

Materials

Student Activity Sheet 3

Prerequisites

Students should have had 20–30 minutes of exploration with *TesselMania!* before doing this activity. They should practice using the tack tool to place points, the scissors tool to remove points, and the arrow tool to move points or the whole tile.

Notes

This activity is provided for use with those students who may find it challenging to create tile shapes with *TesselMania!* In later activities, students need to reproduce tiles on the computer that are given on the activity sheets. Not only does this activity provide practice, but it also encourages students to experiment creatively by making many different shapes with the same number of tack points.

This activity is optional. Many students who have had experience in using software and computer drawing tools will find this extra practice unnecessary.

In Class

Begin with a very short demonstration, using *TesselMania!* Choose the Translation tile. Remind students how to select the tack tool and how to use it to place a point on the top side of the tile. Show how to move the tack point in several directions.

Draw a simple tile side shape on the board. (You can use one of those on the student activity sheet.) Ask students how many tack points you need to add so that you can create this tile side. Place that number of tack points and try to create the shape. Discuss whether or not it works. Be sure students see how to change the length of line segments and how to change the angle.

You may want students to do only Part 1 and then reconvene the class before moving on to Part 2.

In Part 2, students start with just one tack point and create many different shapes by moving that point. They should record three "interesting" shapes on the table in the activity sheet. They can use any criteria they wish to choose which are interesting.

In the Summary, have students compare their table of tile sides with those of other students. The objective here is to notice the wide variety of shapes possible and to begin to relate characteristic shapes with the number of tack points they require.

Creating Tile Sides
Teacher's Notes • *page 2 of 2*

Results

Part 1

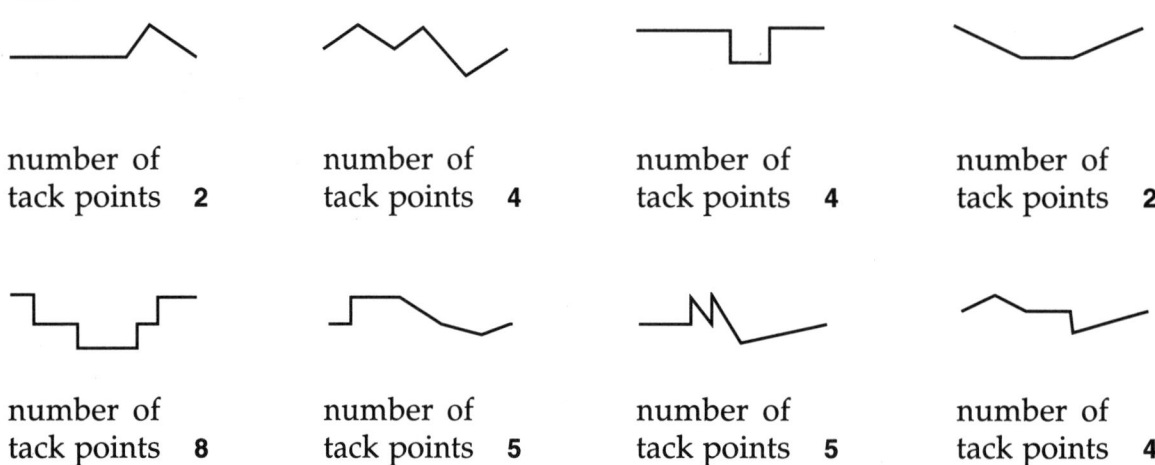

number of tack points   2

number of tack points   4

number of tack points   4

number of tack points   2

number of tack points   8

number of tack points   5

number of tack points   5

number of tack points   4

Part 2

The results in the table of tile sides will vary.

Summary

Write a summary of what you noticed about the shape of a tile side and the number of tack points on it.

Responses will vary.

Challenge

How many tack points do you need to make a curved tile side? Make a conjecture and test it using TesselMania! Refine your conjecture as needed.

Conjectures will vary.

Theoretically speaking, how many points are needed to make a curve? Is this answer the same as the one above? Explain why these two answers are the same or different.

Theoretically, an infinite number of tack points would be needed to make a smooth curve, as opposed to a shape made of very short line segments. The two answers are different because of the way our eyes view the dots of light on a computer screen. We see a smooth curve when the line segments are very short.

Activity 4: Translation and Rotation Tilings

Explore translation and rotation in tessellations, using the Heesch system.

Overview

This is the first of four activities that explore tessellations using the Heesch classification system. You need to know a little about the Heesch system before you begin. If you have not already done so, read the section "Heesch Types" in *TesselMania!* by opening the Options menu and choosing this feature.

Symbols used in Heesch types
T: translation
C: rotation
G: glide reflection

An example is worked for you on this page. Follow each step for this example and then repeat the steps for the problems on the next page. Do Steps a–c on this activity sheet. Do Steps d–g at the computer. Record your results on this sheet.

Steps

a. Study the diagram. Notice that the arrows show transformations of tile sides.
b. Draw the missing sides of the transformed tile.
c. Write how the tile was transformed.
d. Start *TesselMania!* Select the tile type that matches the given tile. Create the tile. Color it.
e. Record the Heesch type of the tile (look in the screen's upper right corner).
f. Observe the tessellation animation (the middle "magic" button). Record the moves used to position the eight adjacent tiles.
g. Select Erase All from the Edit menu. Move the vertices (use the arrow tool). Determine which quadrilaterals can be used for this Heesch type. List them.

Example

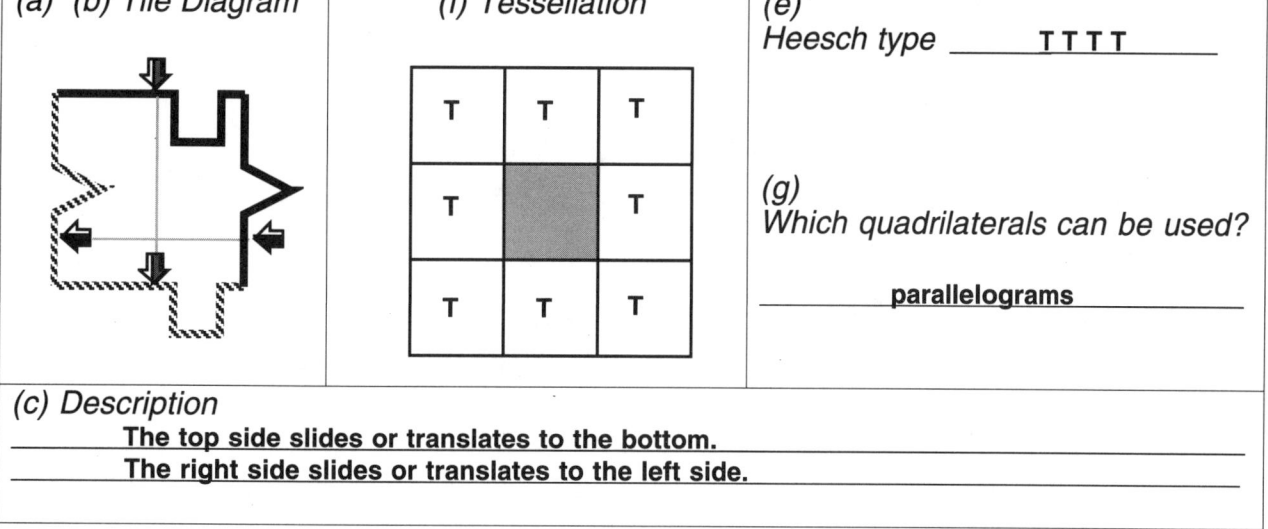

(a) (b) Tile Diagram

(f) Tessellation

(e) Heesch type _____ T T T T _____

(g) Which quadrilaterals can be used?
_____ parallelograms _____

(c) Description
_____ The top side slides or translates to the bottom. _____
_____ The right side slides or translates to the left side. _____

Translation and Rotation Tilings
page 2 of 2

1.

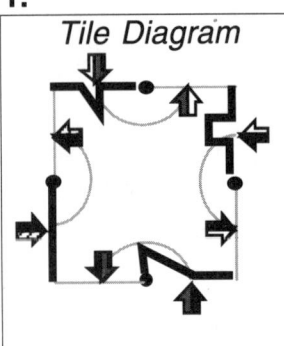
Tile Diagram

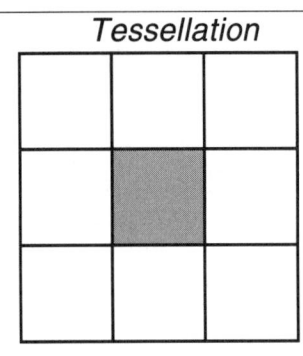
Tessellation

Heesch type _____

Which quadrilaterals can be used?
_____

Description _____
_____

2.

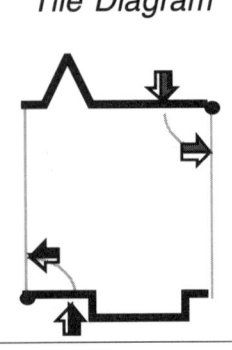
Tile Diagram

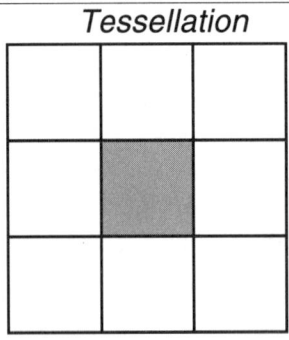
Tessellation

Heesch type _____

Which quadrilaterals can be used?
_____

Description _____
_____

3.

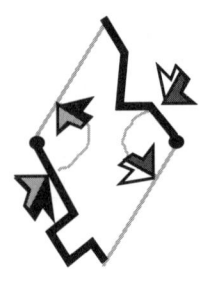
Tile Diagram

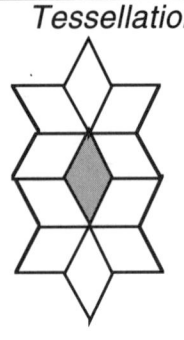
Tessellation

Heesch type _____

Which quadrilaterals can be used?
_____

Description _____
_____

Translation and Rotation Tilings
Activity 4: Teacher's Notes

Objectives

To investigate translations and rotations on quadrilateral tiles
To explore the Heesch system for classifying transformations on tile sides
To relate the Heesch type of a tile to the movements involved in tessellation

Materials

Student Activity Sheet 4
Transparency Master: Activity 4

Prerequisites

Students must be familiar with the *TesselMania!* software. Prior to this activity, students should have had at least 30 minutes to experiment with the software and create imaginative, colorful tessellations. In this activity they will need to focus on data gathering and transformations.

Activity 1 (or a similar hands-on activity that creates tessellations) should be done before this activity so that students will have an intuitive understanding of the three types of transformations involved in tessellations.

Activity 2 is a valuable introduction to the Heesch classification system. In this activity, students must be able to use the tack tool in *TesselMania!* to create tiles of various shapes. Activity 3 provides practice in creating tiles.

Notes

Activities 4, 5, 6, and 7 all involve creating tessellations from the Heesch Types of tiles, as presented in the *TesselMania!* software.

All students should complete Activity 4. Ideally, students should then work on Activities 5, 6, and 7. If time is limited, you may want to have groups complete just one or two of Activities 5, 6, and 7, and then share their results.

Have students work in pairs for this activity. Students must have access to *TesselMania!* If necessary, they can do the tile drawings and descriptions (Steps a–c) without the software, then complete the activity (Steps d–g) when the software is available.

The teacher's demonstration at the beginning of this activity is essential. Be sure you practice this demonstration before class so that you can present it with confidence.

In Class

Explain that students will explore the mathematics of transformations behind the tessellations. Briefly review their previous work with transformations.

Explain that you will work together through the first tessellation, and then the students will work on other, even more interesting tessellations on their own. You will be demonstrating the procedure they will need to follow.

Translation and Rotation Tiles
Teacher's Notes • *page 2 of 4*

Point out that since there are many types of tessellations, it might be useful to *classify* them in order to study and compare them. In the 1930s, mathematician Heinrich Heesch developed a system for classifying asymmetrical, isohedral tessellations. Note that Heesch was a mathematician, and Escher was a graphic artist.

On the demonstration computer, show and read the three screens of information on the Heesch types (under the Options menu). Emphasize that the Heesch types are labeled by moving around the tile clockwise, one side at a time. Point out that the three letters used for the transformations are *T* (translation), *C* (rotation) and *G* (glide reflection). While it might seem more natural to use the letter *R* for rotation, the C indicates the *center* of rotation. Some students prefer to think of the C as *circle*.

On the overhead or board, show the diagram of the tile in the worked example (Transparency Master: Activity 4). Point out the arrows. Draw the missing sides, describing how translations are involved. Check that students understand this process by asking someone to describe it to the class. Point out that they will need to *write* a description of this process on their activity sheets. If students use *slide*, introduce the term *translation*. (They need to know the terms *translation, rotation,* and *glide reflection* in order to use the main menu of the software.)

Move back to the computer. At the Main menu, select Translation because we know that translations are used to create the tile. Click on the Show Me button and observe the animation.

Open the Translation section and choose the quadrilateral tile. Point out the Heesch label at the upper right corner of the drawing window. Create the tile from the Example (or a similar tile). Color it.

Click the middle "magic" button to show the tessellation animation. Adjust the speed control to medium at first, faster if the students find the speed too slow. After all eight adjacent tiles have been placed, click on Pause. Ask students to describe how the tile *moved* to each of the eight adjacent positions (translations right, left, up, down, and four diagonals).

On the overhead or the board, show the 9-tile grid. Show students how to *record* the moves by writing the letter *T* in the tiles. You may want to draw arrows to indicate the direction of the translation.

This process demonstrates that any tessellation based on a *square* will be created with these same moves. What about other quadrilaterals? Back at the computer, select Erase All from the Edit menu. Use the arrow cursor. Click and drag on the open-dot vertices and change the shape of the quadrilateral in various ways. Ask students what kinds of quadrilaterals you are creating. (All are parallelograms.) Does this make sense? To translate one side to the opposite side of a quadrilateral, do the sides need to be parallel? (Yes.) So this demonstrates but does not *prove* that this Heesch type, TTTT, works on any parallelogram.

Ask if students have any questions about this procedure. Explain that this example is shown on the first page of their activity sheet so that they can refer back to it for help if needed. If your students need extra help getting started, they can redo the example problem themselves. Otherwise they can move to page 2 and begin with the tiles

Translation and Rotation Tiles
Teacher's Notes • page 3 of 4

formed by rotation. Encourage students to note and describe the *center of rotation*, the *amount of rotation* (degrees), and the *direction* for each rotation. Emphasize that they are to collect and record data (the transformations) and write their descriptions in sentences.

During the lab work, assist students as needed and notice which concepts are causing difficulty or confusion.

After the activity sheets are completed, hold a class discussion on the results. Ask students to state, and defend if necessary, their results for Heesch type, tessellation moves, and types of quadrilaterals.

Challenge advanced students to find relationships between the Heesch type and the tessellation moves. For example, if translations are involved in the tile, then translations will also be in the moves, but even if only rotations are used in the tile, translations can be part of the moves.

Results
Example

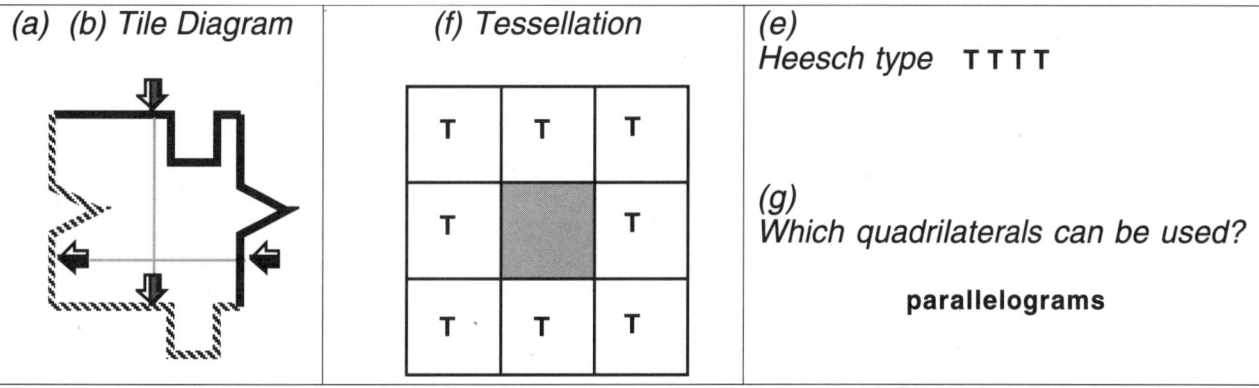

(a) (b) Tile Diagram	(f) Tessellation	(e) Heesch type T T T T
		(g) Which quadrilaterals can be used? **parallelograms**

(c) Description
The top side slides or translates to the bottom.
The right side slides or translates to the left.

1.

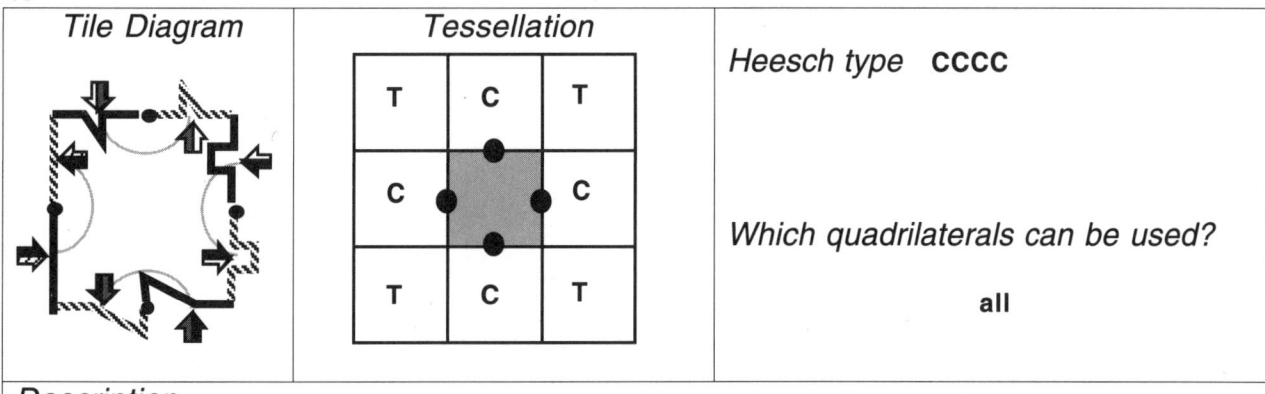

Tile Diagram	Tessellation	Heesch type CCCC
		Which quadrilaterals can be used? **all**

Description
Each of the four sides is rotated around its midpoint. The rotation is 180°.

TesselMania! Math Connection

Translation and Rotation Tilings
Teacher's Notes • page 4 of 4

2.

Tile Diagram	Tessellation	
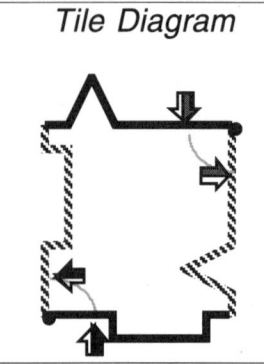	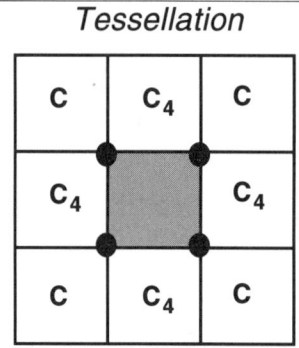	Heesch type $C_4 C_4 C_4 C_4$ Which quadrilaterals can be used? squares (90° angles, equal adjacent sides)

Description
For two opposite sides, the side is rotated 90° counterclockwise around the adjacent corner point (vertex), moving clockwise around the tile.

3.

Tile Diagram	Tessellation	
	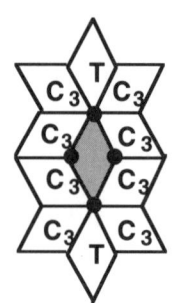	Heesch type $C_3 C_3 C_3 C_3$ Which quadrilaterals can be used? parallelograms with equal sides and 120° angles

Description
For two opposite sides, the side is rotated 120° counterclockwise around the adjacent corner point (vertex), moving clockwise around the tile.

Activity 5: Glide-Reflection Tilings

Explore glide reflections in tessellations, using the Heesch system.

The steps in this activity are the same as in Activity 4.

Steps

a. Study the diagram. Notice that the arrows show transformations of tile sides.
b. Draw the missing sides of the transformed tile.
c. Write how the tile was transformed.
d. At the Main Menu, find the tile type that matches the given tile. Create the tile. Color it.
e. Record the Heesch type of the tile.
f. Observe the tessellation animation. Record the moves for the adjacent tiles.
g. Select Erase All from the Edit menu. Move the vertices. Determine which quadrilaterals can be used for this Heesch type. List them.

1.

Tile Diagram | Tessellation

Heesch type _____

Which quadrilaterals can be used?

_____

Description _____

2.

Tile Diagram | Tessellation

Heesch type _____

Which quadrilaterals?

_____

Description _____

TesselMania! Math Connection

Glide-Reflection Tilings

Activity 5: Teacher's Notes

Objectives

To investigate glide reflections on quadrilateral tiles
To explore the Heesch Types for classifying transformations on tile sides
To relate the Heesch Types to the movements involved in tessellations

Materials

Student Activity Sheet 5
Transparency Master: Activities 5 and 6

Prerequisites

Students should have completed Activity 4.

Notes

Activities 4, 5, 6, and 7 all involve creating tessellations from the Heesch types of tiles, as presented in the *TesselMania!* software.

All students should complete Activity 4. Ideally, students should then work on Activities 5, 6, and 7. If time is limited, you may want to have groups complete just one or two of Activities 5, 6, and 7, and then share their results.

Have students work in pairs for this activity. Students must have access to *TesselMania!* If necessary, they can do the tile drawings and descriptions without the software and then complete the activity when the software is available.

In Class

Students have already explored *TesselMania!* and completed Activity 4. Explain that this activity is similar to Activity 4, but focuses on glide reflections. Students should follow the same steps they used in Activity 4.

During the lab work, assist students as needed and notice which concepts are causing difficulty or confusion.

After the activity sheets are completed, hold a class discussion on the results. Ask students to state, and defend if necessary, their results for Heesch type, tessellation moves, and types of quadrilaterals.

Challenge advanced students to find relationships between the Heesch type and the tessellation moves. For example, if glide reflections are involved in the tile, then glide reflections will also be in the moves, but other transformations may be used in the moves also.

Glide-Reflection Tilings

Teacher's Notes

Results

1.

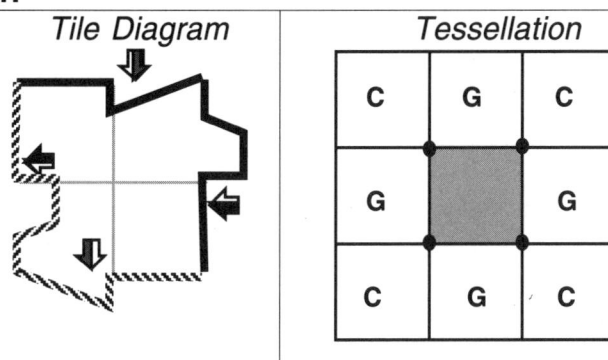

	Tile Diagram	Tessellation	
		C G C	Heesch type $G_1 G_2 G_1 G_2$
		G ▨ G	
		C G C	Which quadrilaterals can be used? **rectangles**

Description
The top is glide-reflected to the bottom across a vertical line of reflection.
The right is glide-reflected to the left across a horizontal line of reflection.

2.

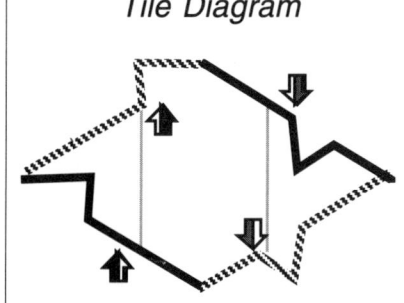
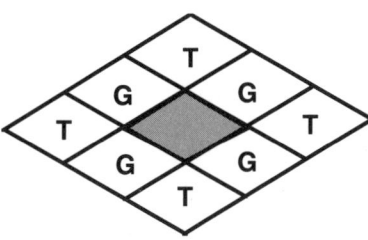

Heesch type $G_1 G_1 G_2 G_2$

Which quadrilaterals?
**kites
(two pairs of equal
adjacent sides)**

Description
One side is glide-reflected across a vertical line to its adjacent side.
The opposite side is glide-reflected across a vertical line to its adjacent side.

Activity 6: Combination Tilings

Explore combinations of transformations in tessellations, using the Heesch system.

The steps in this activity are the same as in Activities 4 and 5.

Steps
- a. Study the diagram. Notice that the arrows show transformations of tile sides.
- b. Draw the missing sides of the transformed tile.
- c. Write how the tile was transformed.
- d. At the Main Menu, find the tile type that matches the given tile. Create the tile. Color it.
- e. Record the Heesch type of the tile.
- f. Observe the tessellation animation. Record the moves for the adjacent tiles.
- g. Select Erase All from the Edit menu. Move the vertices. Determine which quadrilaterals can be used for this Heesch type. List them.

1.

Diagram | Tessellation | Heesch type _____

Which quadrilaterals can be used? _____

Description _____

2.

Diagram | Tessellation | Heesch type _____

Which quadrilaterals can be used? _____

Description _____

TesselMania! Math Connection — ©1995 by Key Curriculum Press

Combination Tilings
page 2 of 2

3.

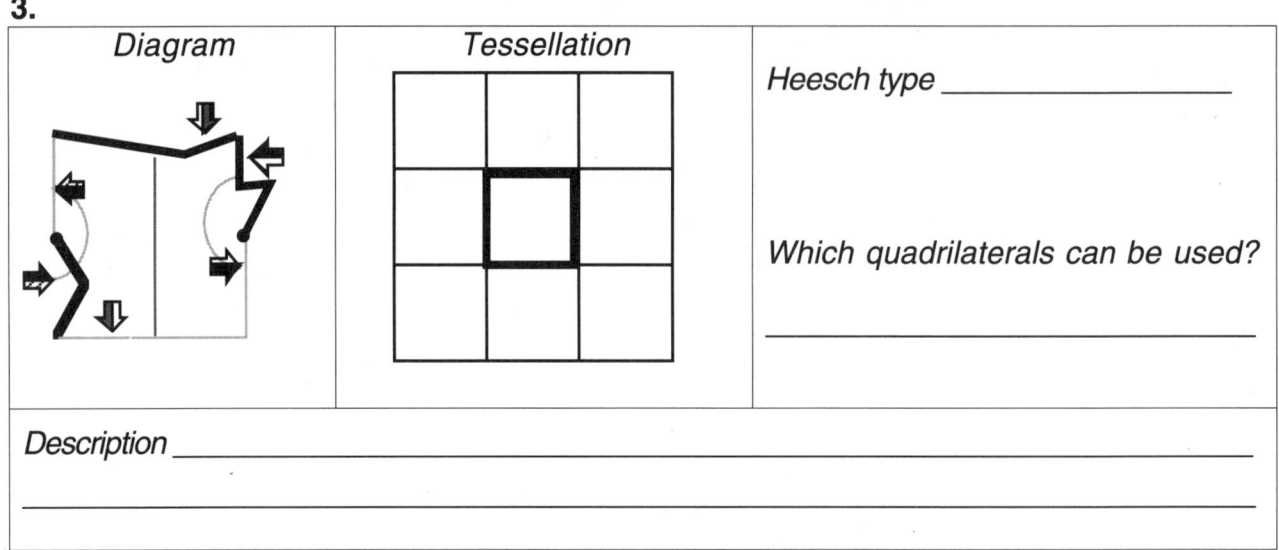

Diagram	Tessellation	
		Heesch type _____
		Which quadrilaterals can be used? _____

Description _____
_____

4.

Diagram	Tessellation	
		Heesch type _____
		Which quadrilaterals can be used? _____

Description _____
_____

Challenge

Why aren't the patterns TTCC and TTGG included?

_____
_____
_____
_____

Combination Tilings

Activity 6: Teacher's Notes

Objectives

To investigate combinations of transformations on quadrilateral tiles
To explore the Heesch types for classifying transformations on tile sides
To relate the Heesch types to the movements involved in tessellations

Materials

Student Activity Sheet 6
Transparency Master: Activities 5 and 6

Prerequisites

Students should have completed Activity Sheet 4.

Notes

Activities 4, 5, 6, and 7 all involve creating tessellations from the Heesch types of tiles, as presented in the *TesselMania!* software. All students should complete Activity 4. Ideally, students should also complete Activities 5, 6, and 7.

Students must have access to *TesselMania!* during this activity. If necessary, they can do the tile drawings and descriptions without the software and then complete the activity when the software is available.

In Class

Explain that this activity is similar to Activities 4 and 5, but focuses on combinations of transformations. Students should follow the same steps they used in Activity 4. During the lab work, assist students as needed and notice which concepts are causing difficulty or confusion.

After the activity sheets are completed, hold a class discussion on the results. Ask students to state, and defend if necessary, their results for Heesch type, tessellation moves, and types of quadrilaterals. Challenge advanced students to find relationships between the Heesch type and the tessellation moves.

Results

1.

Tile Diagram	Tessellation	
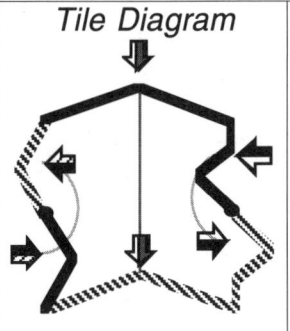	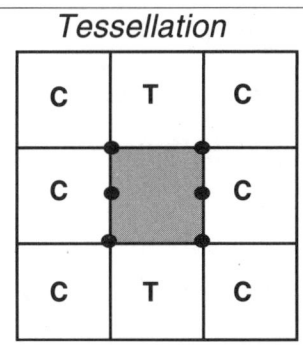	*Heesch type* **TCTC** *Which quadrilaterals can be used?* **parallelograms**

Description
The top is translated to the bottom. The sides have 180° rotations around their centers.

Combination Tilings
Teacher's Notes • page 2 of 2

2.

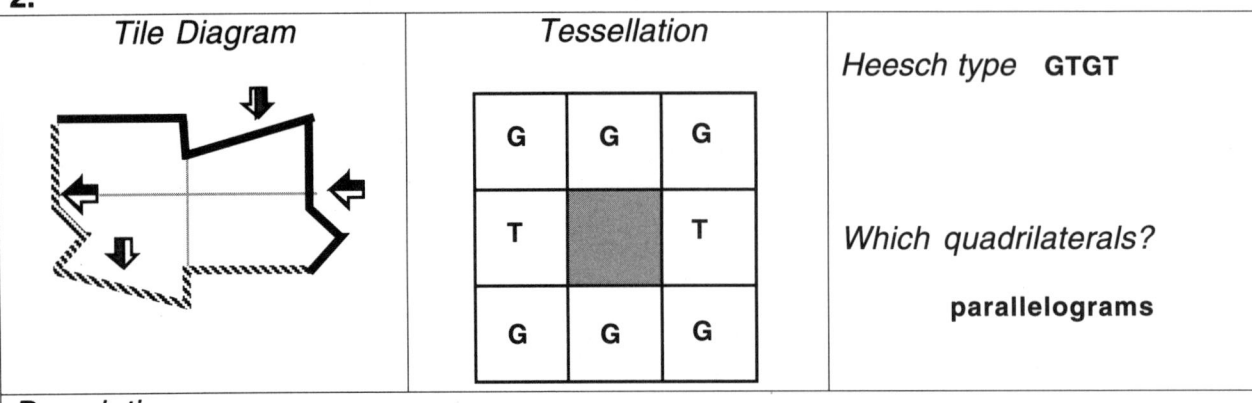

Heesch type **GTGT**

Which quadrilaterals?

    parallelograms

Description
The top is glide-reflected over a vertical line to the opposite side (bottom). The right side is translated to the left side.

3.

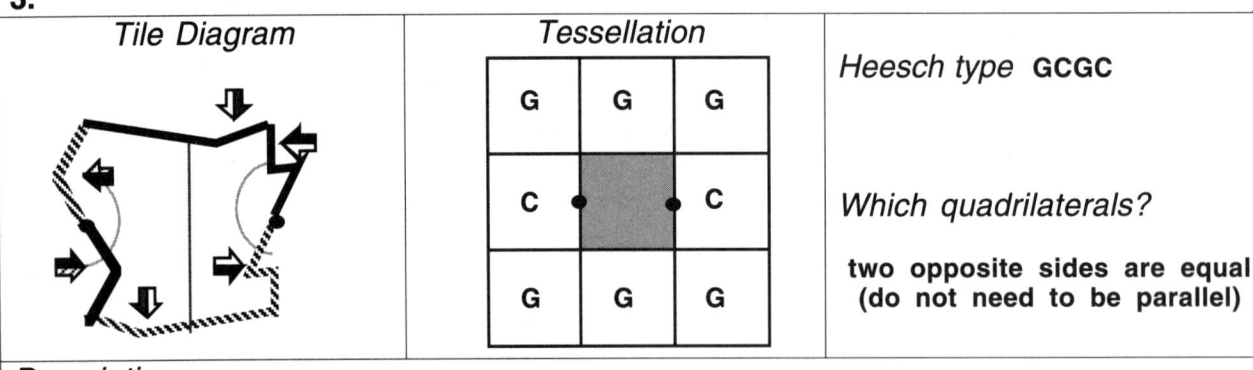

Heesch type **GCGC**

Which quadrilaterals?

    two opposite sides are equal (do not need to be parallel)

Description
One side is glide-reflected over a vertical line to the opposite side. Each of the other two sides has a 180° rotation around its center.

4.

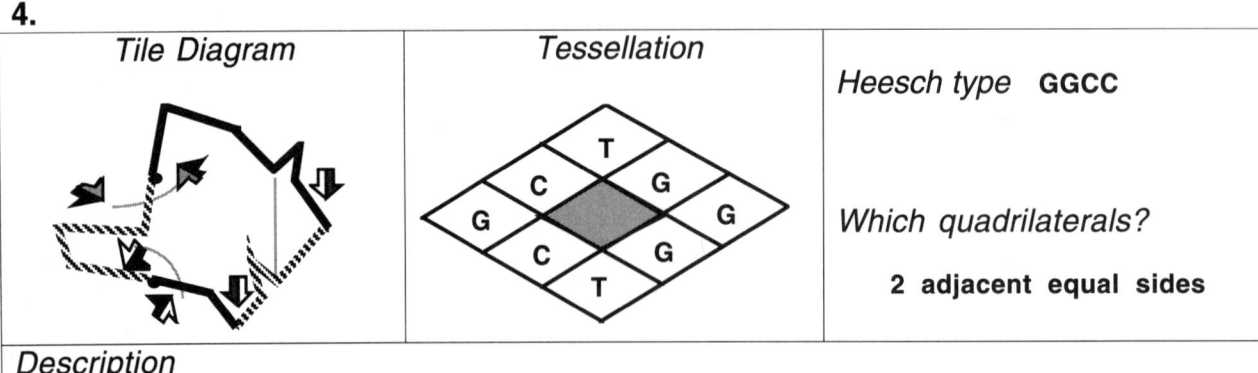

Heesch type **GGCC**

Which quadrilaterals?

    2 adjacent equal sides

Description
One side is glide-reflected over a vertical line to the adjacent side. The other two sides have 180° rotations around their centers.

Challenge

Why aren't the patterns TTCC and TTGG included?

TTCC implies that the first side is translated, but the third (opposite) side is rotated. This is not possible. The same reasoning applies to TTGG.

Activity 7: Triangle and Hexagon Tilings

Explore triangle and hexagon tessellations, using the Heesch system.

The steps in this activity are the same as in Activities 4, 5, and 6.

Steps

a. Study the diagram. Notice that the arrows show transformations of tile sides.
b. Draw the missing sides of the transformed tile.
c. Write how the tile was transformed.
d. At the Main Menu, find the tile type that matches the given tile. Create the tile. Color it.
e. Record the Heesch type of the tile.
f. Observe the tessellation animation. Record the moves for the adjacent tiles.
g. Select Erase All from the Edit menu. Move the vertices.
   Determine which quadrilaterals can be used for this Heesch type. List them.

1.

Diagram   Tessellation

Heesch type _____

Which triangles can be used?
_____

Description _____
_____

2.

Diagram   Tessellation

Heesch type _____

Which triangles can be used?
_____

Description _____
_____

TesselMania! Math Connection   Activity 7 • 43

Triangle and Hexagon Tilings

page 2 of 2

3.

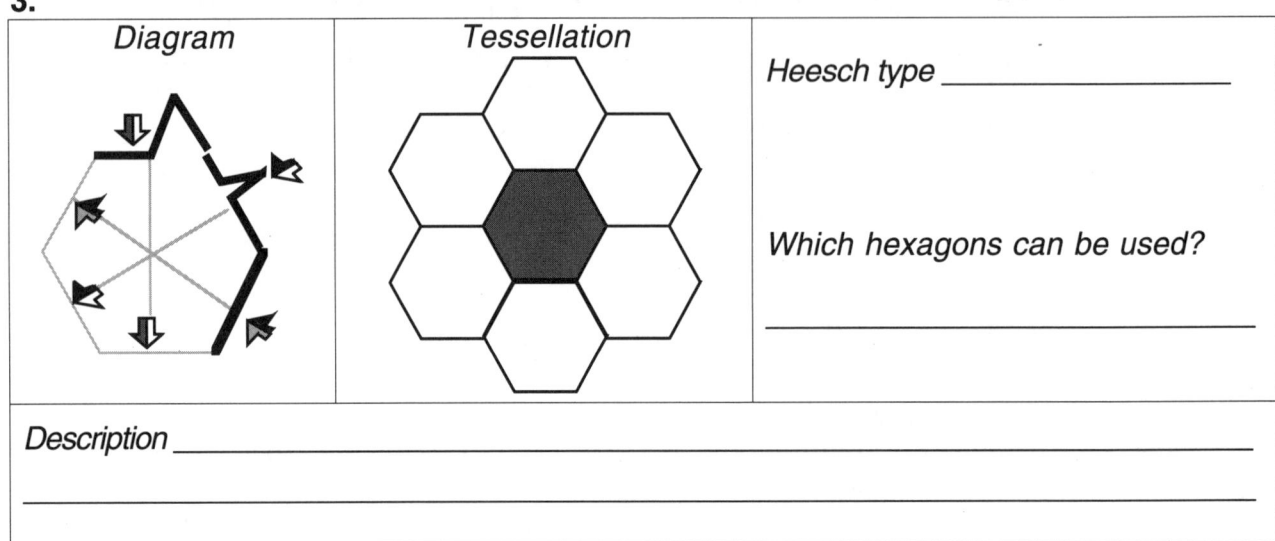

Diagram | Tessellation

Heesch type _____

Which hexagons can be used?

_____

Description _____

_____

4.

Diagram | Tessellation

Heesch type _____

Which hexagons can be used?

_____

Description _____

_____

Challenge

Why isn't the pattern TTC included?

_____

What other transformations (that are not possible to create on *TesselMania!*) can you find that tessellate for triangles or hexagons?

_____

_____

Triangle and Hexagon Tilings
Activity 7:   Teacher's Notes

Objectives

To investigate transformations on triangle and hexagon tiles
To explore the Heesch types for classifying transformations on tile sides
To relate the Heesch types to the movements involved in tessellations

Materials

Student Activity Sheet 7
Transparency Master: Activity 7

Prerequisites

Students should have completed Activities 4, 5, and 6.

Notes

Activities 4, 5, 6, and 7 all involve creating tessellations from the Heesch types of tiles, as presented in the *TesselMania!* software. All students should complete Activity 4. Ideally, students should then work on Activities 5, 6, and 7.

Students must have access to *TesselMania!* during this activity. If necessary, they can do the tile drawings and descriptions without the software and then complete the activity when the software is available.

In Class

Explain that this activity is similar to Activity 4, but focuses on triangles and hexagons instead of quadrilaterals. Students should follow the same steps they used in Activity 4. During the lab work, assist students as needed and notice which concepts are causing difficulty or confusion.

After the activity sheets are completed, hold a class discussion on the results. Ask students to state, and defend if necessary, their results for Heesch type, tessellation moves, and types of shapes. Challenge advanced students to find relationships between the Heesch type and the tessellation moves.

Results

1.

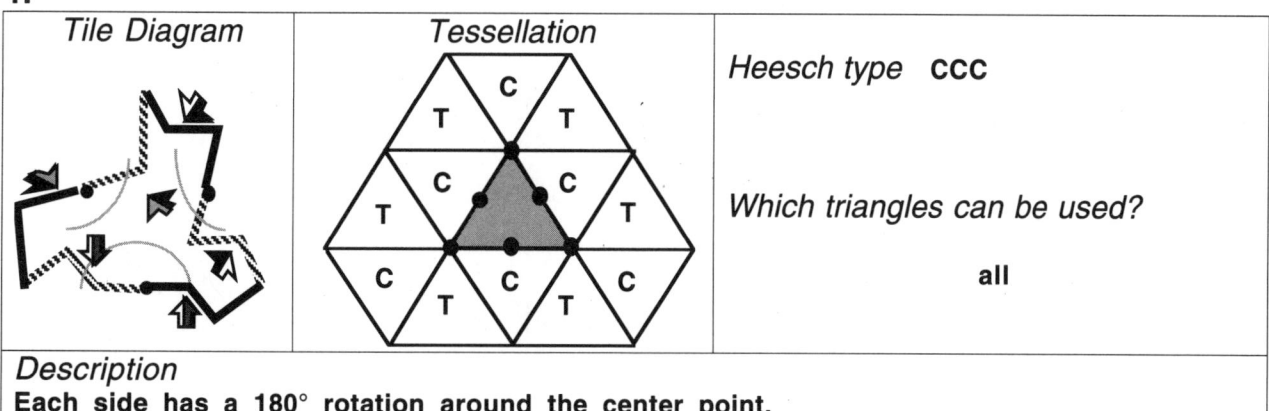

Tile Diagram	Tessellation	
		Heesch type **CCC**
		Which triangles can be used?
		all
Description		
Each side has a 180° rotation around the center point.		

Triangle and Hexagon Tilings
Teacher's Notes • *page 2 of 3*

2.

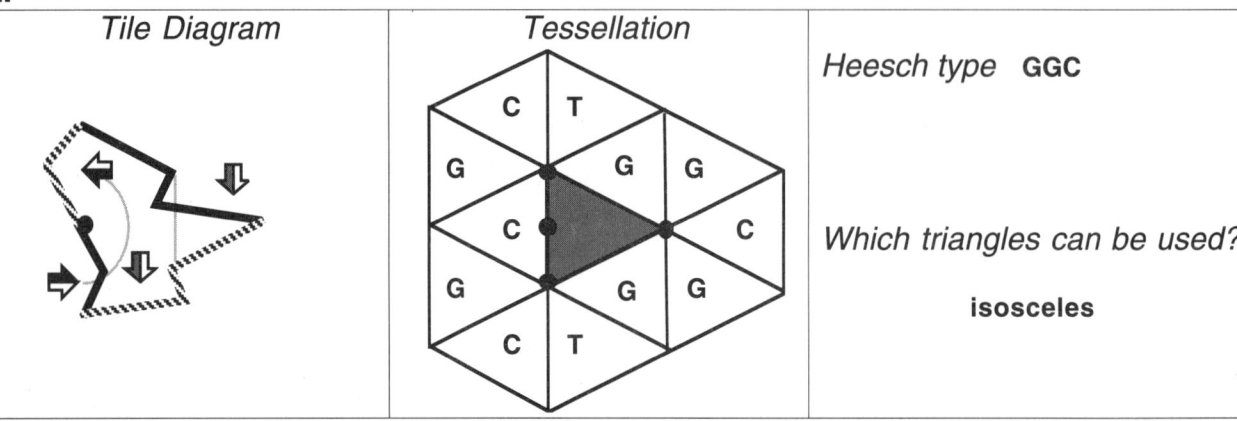

Tile Diagram | Tessellation

Heesch type **GGC**

Which triangles can be used?

isosceles

Description
One of the two equal sides is glide-reflected to the other. The third side has a rotation of 180° around the center point.

3.

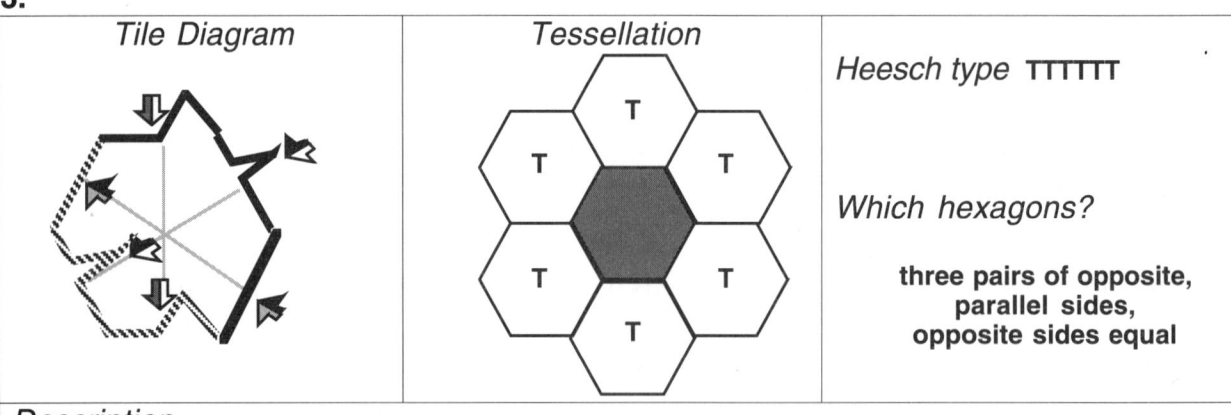

Tile Diagram | Tessellation

Heesch type **TTTTTT**

Which hexagons?

three pairs of opposite, parallel sides, opposite sides equal

Description
The top side translates to the bottom. The right upper side translates to the left lower side. The right lower side translates to the left upper side.

4.

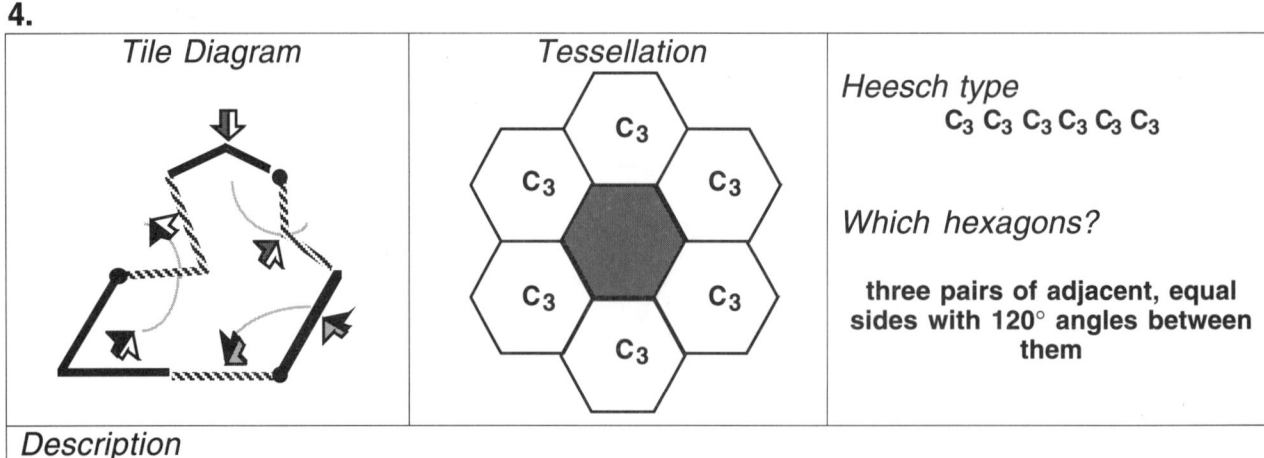

Tile Diagram | Tessellation

Heesch type
$C_3\ C_3\ C_3\ C_3\ C_3\ C_3$

Which hexagons?

three pairs of adjacent, equal sides with 120° angles between them

Description
Every other side is rotated 120° around a vertex to the adjacent side.

Triangle and Hexagon Tilings
Teacher's Notes • page 3 of 3

Challenge

Why isn't the pattern TTC included?

TTC is not possible, because translation requires opposite, parallel sides, which are not possible in a triangle.

What other transformations (that are not possible to create on TesselMania!) can you find that tessellate for triangles or hexagons?

Triangles	Hexagons
C C_6 C_6	T C C T C C
C C_4 C_4	T G_1 G_1 T G_2 G_2
C C_3 C_3	T G_1 G_2 T G_2 G_1
	T C C T G G
	C G_1 C G_2 G_1 G_2

Activity 8: Escher's Classification System

Investigate the Escher system of classifying tessellations.

Overview

The graphic artist M. C. Escher (1898–1972) created his own method of classifying the tessellations he created. He divided the tessellations into ten principal systems. Each system describes how the tessellation is created from translations, rotations, and glide reflections.

In this activity you will analyze examples of tessellations for each of the nine Escher systems that use quadrilaterals. By recording your findings in the table on the next page, you will create a classification system like Escher's. You will then compare this system with the Heesch system.

Directions

- Start *TesselMania!* Select Open. Open the *TesselMania!* Examples folder.
- Open one of the Example Tessellations listed in the table on the next page.
- Use the middle "magic" button to observe the movements in the tessellation.
- Identify the glide reflections and the translations as *transversal* or *diagonal* (see the definitions illustrated below).

    Record the *number* of translations or glide reflections on the table.

- Identify the rotations, the centers of rotation, and the degrees of rotation.

    Record the number and type of rotations on the table.

- Write the Heesch type in the first column of the table.
- Continue in the same way with the other example tessellations.

Transversal is a direction parallel to a tile side.
Diagonal is the direction of a line segment that connects opposite vertices.

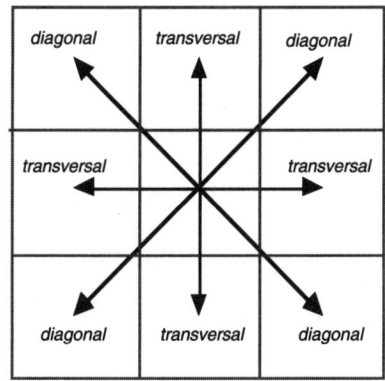
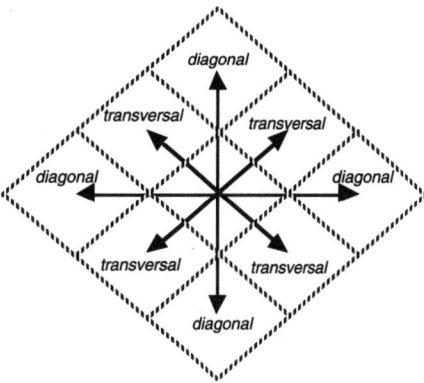

TesselMania! Math Connection

Escher's Classification System

Classification Chart

Heesch Type	Escher System	Example Tessellation	Translations	Rotations	Glide Reflections
	I	Cats	transversal _____ diagonal _____		transversal _____ diagonal _____
	II	Duck	transversal _____ diagonal _____		transversal _____ diagonal _____
	III	Eagles	transversal _____ diagonal _____		transversal _____ diagonal _____
	IV	Lizards	transversal _____ diagonal _____		transversal _____ diagonal _____
	V	Fish	transversal _____ diagonal _____		transversal _____ diagonal _____
	VI	Turtle	transversal _____ diagonal _____		transversal _____ diagonal _____
	VII	Winking cat	transversal _____ diagonal _____		transversal _____ diagonal _____
	VIII	Mushrooms	transversal _____ diagonal _____		transversal _____ diagonal _____
	IX	Fish girl	transversal _____ diagonal _____		transversal _____ diagonal _____

Analysis

Is there a relationship between the Heesch types and the Escher system? If so, describe it. If not, explain why.

_____
_____
_____
_____

Evaluation

Which system of classification do you prefer? Explain your reasons.

_____
_____
_____

Escher's Classification System

Activity 8: Teacher's Notes

Objective

To investigate the Escher system of classifying tessellations
To compare the Escher system with the Heesch system

Materials

Student Activity Sheet 8
The *TesselMania!* Examples folder on the *TesselMania!* disk
Transparency Master: Activity 8

Prerequisites

Students must have completed Activity 4 and at least some of Activities 5, 6, and 7. They need to understand how to describe the tile movements that create a tessellation. They need to have had experience in using the Heesch system to compare that system with Escher's.

Background

The graphic artist M. C. Escher (1898–1972) created his own system of classifying the tessellations he created. He divided the tessellations into ten principal systems. Each system describes how the tessellation is created from translations, rotations, and glide reflections.

This activity explores the first nine systems (I–IX), which are those that use quadrilaterals. System X uses triangles. Escher's classification scheme is actually more complex than the version presented in this activity. For a detailed description of Escher's classification system and tables that classify his drawings by system, see *Visions of Symmetry* by Doris Schattschneider. A reproduction of Escher's classification table is included on page 61. Her book also relates the Heesch types to the Escher systems.

In Class

It is essential to begin this activity with a teacher-led demonstration. You may need to do more than one example with the class. Start with the simplest system, System I. Use the Cats file from the *TesselMania!* Examples folder. Follow the directions in the student worksheet. Emphasize that you are following the same steps the students will follow on their own.

Be sure students understand the meaning of the terms *transversal* and *diagonal* as used by Escher. Use Transparency Master: Activity 8 to point out that the transversal direction is *parallel to a tile side*.

Be sure that students understand the columns on the table and how to record the information. They should write the number of transformations, describe the rotations as being on the centers or the vertices, and state the number of degrees of rotation.

Escher's Classification System
Teacher's Notes • page 2 of 4

You may want to do a second example with the class that involves more complicated transformations. You can choose one of the other tessellation files listed in the Classification Chart or one from the table below. A complete list of all the files in the *TesselMania!* Examples folder is included at the end of this section.

Escher System	Heesch Type	Tessellation
II	TCTC	Fall leaves, Elf guy, Hag, Gnomes, Tragedy/comedy
IV	$G_1 G_1 G_2 G_2$	Rose sample
V	GTGT	Cupid
VI	GGCC	Sharkey, Goldfish 2

Results

Classification Chart

Heesch Type	Escher System	Example Tessellations	Translations		Rotations	Glide Reflections	
TTTT	I	Cats	transversal	4			
			diagonal	4			
TCTC	II	Duck	transversal	2	4 (180°) on vertices, 2 (90°) on centers		
			diagonal	0			
CCCC	III	Eagles	transversal	0	4 (180°) on centers		
			diagonal	4			
$G_1 G_1$-$G_2 G_2$	IV	Lizards	transversal	0		transversal	4
			diagonal	4		diagonal	0
GTGT	V	Fish	transversal	2		transversal	2
			diagonal	0		diagonal	4
GGCC	VI	Turtle	transversal	0	2 (180°) on centers	transversal	2 *
			diagonal	2		diagonal	2
GCGC	VII	Winking cat			2 (180°) on centers	transversal	2
						diagonal	4
$G_1 G_2$-$G_1 G_2$	VIII	Mushrooms			4 (180°) on vertices	transversal	4
						diagonal	0
$C_4 C_4$-$C_4 C_4$	IX	Fish girl			8 on diagonal vertices (90° and 180°)		

*This tessellation has glide reflections in both transversal directions, but only in the direction of the sides without a rotation point.

Escher's Classification System

Teacher's Notes • page 3 of 4

Analysis

Is there a relationship between the Heesch types and the Escher system? If so, describe it. If not, explain why.

Yes, there seems to be a relationship between the Heesch types and the Escher system. Each of the Escher systems (I–IX) matches one of the Heesch types. However, there are at least 11 Heesch types in *TesselMania!*, and we don't know about matching the other two Heesch types to the Escher system.

Some Heesch types use triangles and hexagons. The table for the Escher system shows only quadrilaterals.

Evaluation

Which system of classification do you prefer? Explain your reasons.

Student responses will vary.

Most people find the Heesch system easier to understand and to use. That is why it is used in the *TesselMania!* software. The Heesch system is based on an individual tile rather than the tessellation. It seems easier to identify the transformations of the tile sides than to identify those used to make the tessellation.

Heesch created his system to be published and used by other people. Escher created his system strictly for his own personal use in creating and organizing his artwork. The fact that Heesch was a mathematician and Escher was not may also have had some influence on the systems each constructed.

Alternate Approach or Extension

In this activity, students reconstructed parts of Escher's classification system by analyzing known examples of Escher's systems. Another approach to studying the Escher system is to give students the completed table and a collection of tessellations. Ask students to identify which of the Escher systems each tessellation represents.

This second approach requires that students understand the classification table. The original table, which is found in *Visions of Symmetry*, looks complex and contains language that students may find difficult (e.g., "2-fold" instead of 180°).

You may want to try this approach after students have completed Activity 8. You can show a tessellation example to the whole class and ask them to analyze it and classify it. You may also want to have students classify other tessellations that are included on the *TesselMania!* disk, that are reprinted in textbooks, or that other students create.

Escher's Classification System

Teacher's Notes • page 4 of 4

Contents of *TesselMania!* Examples Folder on the *TesselMania!* Disk

File Name	Tile Shape (Quadrilateral unless marked)	Heesch Type	Escher System
Birds	hexagon	TTTTTT	
Bunny		$C_3 C_3 C_3 C_3$	
Butterflies		$C_3 C_3 C_6 C_6$	
Cats		TTTT	I
Cornucopia	hexagon	$C_3 C_3 C_3 C_3 C_3 C_3$	
Cupids		GTGT	V
Duck		TCTC	II
Eagles		CCCC	III
Elf guy		TCTC	II
Fall leaves		TCTC	II
Fish		GTGT	V
Fish girl		$C_4 C_4 C_4 C_4$	IX
Girl		$C_3 C_3 C_3 C_3$	
Gnomes		TCTC	II
Goldfish 2		GGCC	VI
Hag		TCTC	II
Lizard in suit		TTTT	I
Lizards		$G_1 G_1 G_2 G_2$	IV
Mountain fish	triangle	CCC	
Mushrooms		$G_1 G_2 G_1 G_2$	VIII
Parker 1		TTTT	I
Penguins 2		$C_3 C_3 C_3 C_3$	
Picasso-like		$C_3 C_3 C_6 C_6$	
Pigs and corn	triangle	GGC	
Rose sample		$G_1 G_1 G_2 G_2$	IV
Rose window		TTTT	I
Sharkey		GGCC	VI
Squirrels	hexagon	GTGTGTGT	
The rose is fairest		$C_3 C_3 C_3 C_3$	
Tilosaurus tessalee		TTTT	I
Tragedy/comedy		TCTC	II
Turtle		GGCC	VI
Winking cat		GCGC	VII

Note: Some of the tessellations listed above are not designated by an Escher system, because this would require using more complicated notation than has been introduced in this activity.

Activity 9
Reverse Engineering

Analyze and re-create tessellations.

Overview

In this activity, you can use *TesselMania!* to *reverse-engineer* tessellation drawings, working backward from a completed tessellation. In the process, you'll discover its Heesch type and the transformations needed to shape the tile sides. Then you re-create the tessellation, using *TesselMania!* Follow the steps carefully as you re-create the tessellation shown below.

Steps

1. Identify the *motif* (shaped tile) of the tessellation. Trace it onto transparency film.
2. Find the vertices of the original tile (a quadrilateral or a triangle).
3. Trace the original tile outline onto the motif traced on the transparency.
4. Identify how each side is created (translation, rotation, glide reflection).
5. Use the Heesch type symbols (T, C, G) to write the Heesch type of this tile.
6. Start *TesselMania!* Select the Heesch type you found in Step 5.
7. Tape the transparency containing the traced tile onto the computer screen.
8. Use the arrow tool to adjust the shape on the screen to match the original tile.
9. Use the tack tool to change the sides to match the traced motif.
10. Tessellate and check that your tile looks right. Adjust it if needed.
11. Color your tile and add interior artwork.
12. Tessellate and decide whether you want to make more changes. Save your work.

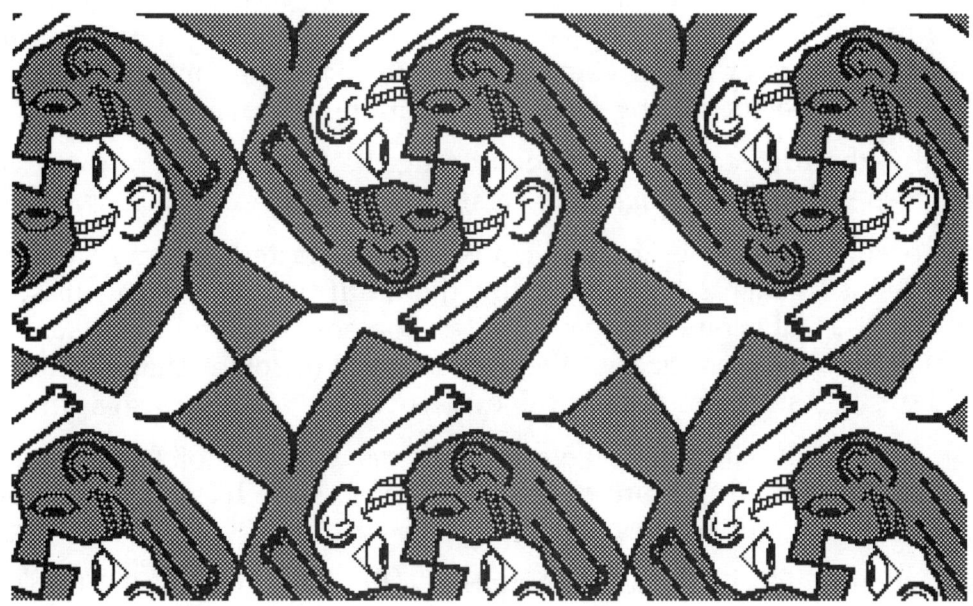

Challenge

Reverse-engineer other tessellations.

TesselMania! Math Connection

Reverse Engineering
Activity 9: Teacher's Notes

Objective

To analyze and re-create tessellations using the Heesch type and *TesselMania!*

Material Needed

Overhead transparency sheets (one quarter of a sheet for each student pair)
Overhead marker pens, one for each student pair
Transparency Masters: Activities 9a and 9b
Student Activity Sheet 9

Prerequisites

Students must understand and have experience using the Heesch classification system and the *TesselMania!* software. They should have completed Activities 4–7.

Notes

This activity begins with a teacher-led demonstration of the process used to reverse-engineer a tessellation. You will want to try it yourself before class to make sure you are comfortable with each of the steps. This will also help you foresee any potential difficulties for your students and enable you to explain the process clearly during class. As you read through the directions for the demonstration, refer to the illustrations on the following two pages.

In Class

Introduce this activity by noting that since students know the Heesch classification system, they can use the Heesch type to classify and reverse-engineer tessellation drawings. You may want to define *reverse engineering* if your students are not familiar with this term. Reverse engineering is the process of analyzing something to learn the details of its design or construction in order to copy or to improve it. You might compare it to the problem-solving strategy of working backward.

In this activity, the first step involves identifying the Heesch type of the tessellation. The second step uses *TesselMania!* to re-create the tile shape.

On the overhead, show the top half of Transparency Master: Activity 9a, the fish tessellation. Ask students to try to visualize the motif that creates this tessellation. Ask about the original geometric shape (this is sometimes called the *fundamental region*). Where would the vertices of the quadrilateral (or other shape) be located? Someone may suggest the point near the eye where the adjacent fish's fin touches.

On the fish tessellation, draw the suggested vertices and sides of the original tile. If students have not noticed it, point out that the vertices are located on the motif at points where more than two tiles meet.

The fundamental region is a quadrilateral. Show the lower half of the transparency, the quadrilateral superimposed on the motif.

Reverse Engineering
Teacher's Notes • page 2 of 3

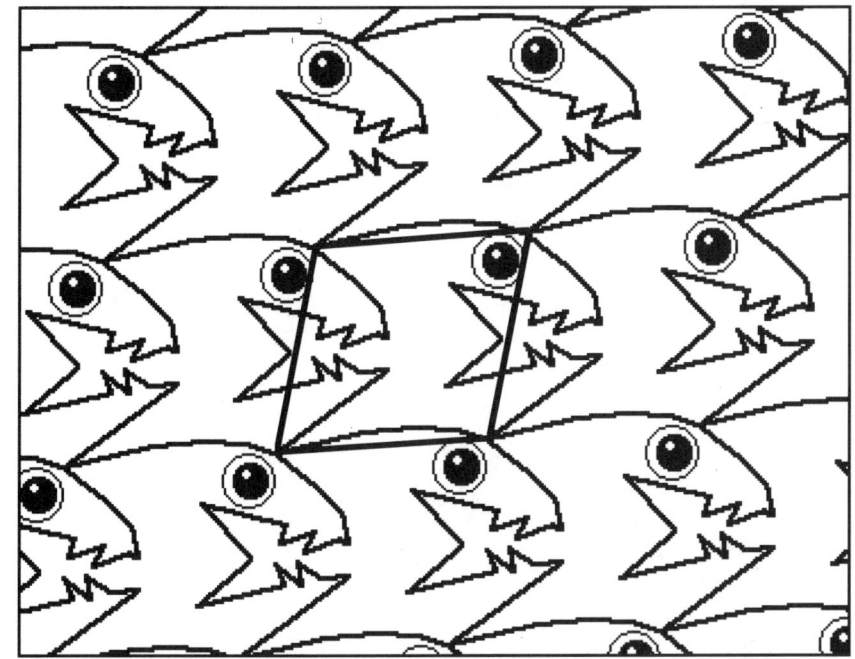
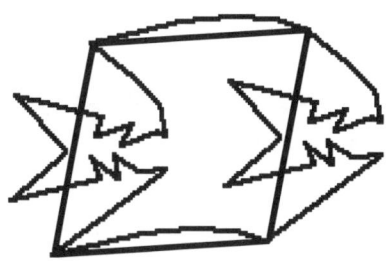

Now, how can we find the Heesch type? First, we know that since the fundamental region is a quadrilateral, the Heesch type must have four letters.

To find the letters of the Heesch type, we need to determine the relationships of the sides. Show Transparency Master: Activity 9b. In this case it is easy to see that the top side is related to the bottom side by translation, so the top and bottom are marked with a T. The left side is related to the right side by translation, so they each can be marked with a T. Putting it all together, we get a Heesch type of TTTT.

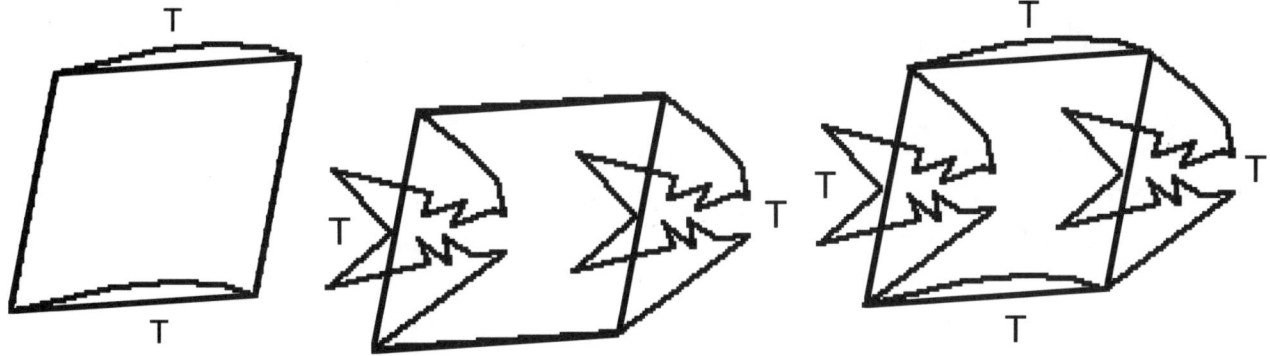

Here's an easy way to duplicate the motif using *TesselMania!:* Trace the original tile (the square) and the motif (the fish) onto a small piece of transparency film.

Start *TesselMania!* and select the TTTT tile type. Tape or hold the transparency tracing onto the computer screen. Use the arrow tool to adjust the square on the screen to match the transparency. (Be sure to look at the computer screen straight on, not at an angle.) Use the tack tool to "bump" the sides of the tile to match the sides of the fish motif.

Reverse Engineering
Teacher's Notes • page 3 of 3

After you have created the fish outline, use the paint tools to duplicate the interior art. Create the tessellation and check that it looks like the original.

Tell students that now they will follow this same process with another tessellation, which involves rotations and a cartoon character. Distribute Activity Sheet 9 and have students work in pairs. Point out that when you re-created the fish tessellation, you followed the same steps as are listed on their activity sheet.

After students have successfully completed this activity, you may want to provide copies of other tessellations, such as those included in this book and the *TesselMania!* manual.

Results

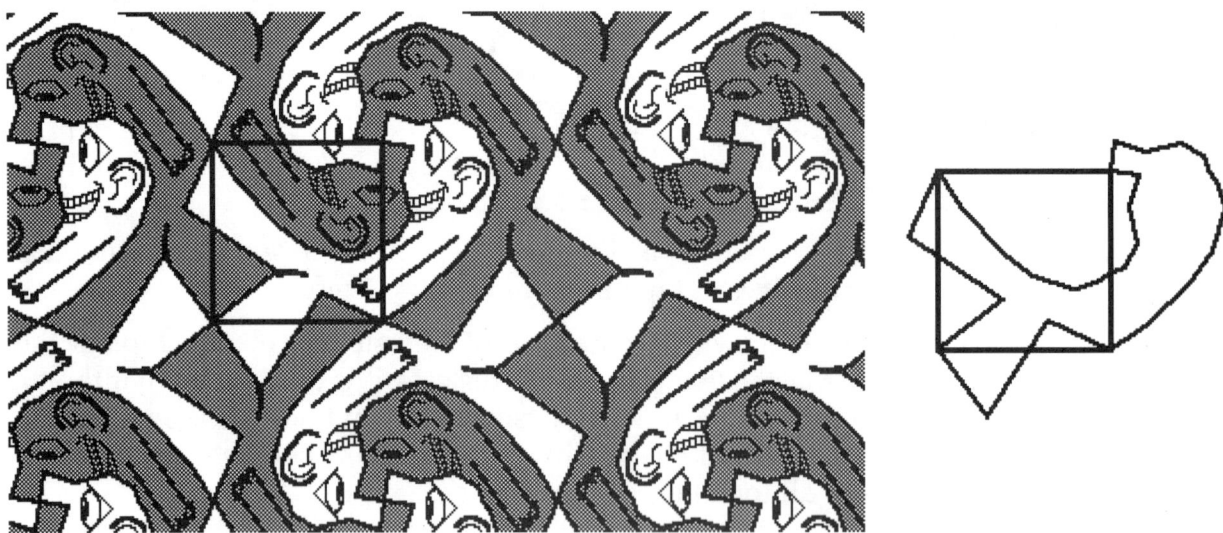

The fundamental region is a square, but the sides are formed by rotations. The sides are rotated 90° around a vertex. Ninety-degree rotation is denoted by a C_4 Heesch label (4-fold center of symmetry, 360 ÷ 4 = 90). The Heesch type is $C_4 C_4 C_4 C_4$.

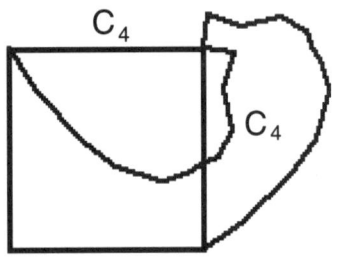 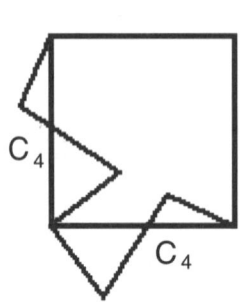 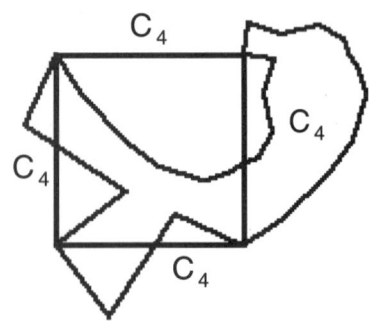

Symmetry Drawing E 106

© M. C. Escher/Cordon Art • Baarn • Holland. All Rights Reserved.

Symmetry Drawing E 104

© M. C. Escher/Cordon Art • Baarn • Holland. All Rights Reserv

Symmetry Drawing E 97

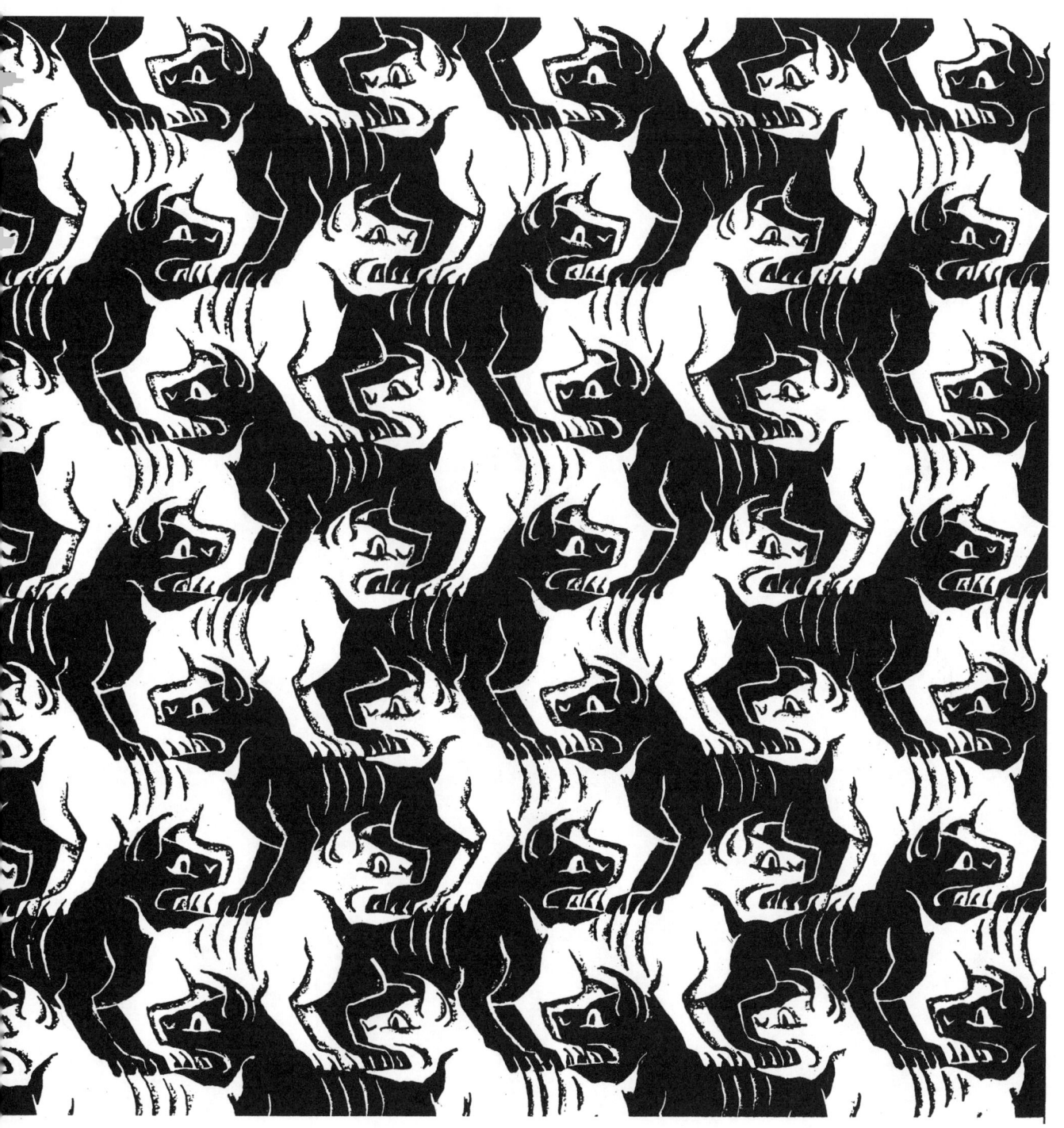

© M. C. Escher/Cordon Art • Baarn • Holland. All Rights Reserved.

Activity 10: Minimal Coloring

Find the minimum number of colors needed to color tessellations.

Overview

Graphic artist M. C. Escher (1898–1972) colored his tessellation drawings so that tiles sharing a common border had contrasting colors. How many colors are needed to accomplish this? In this activity you examine each of the Heesch types of tessellations and determine the smallest number of colors needed to color each type. You will make a conjecture that relates the minimum number of colors to the type of tessellation.

Part 1

Color the two tessellations below so that no two tiles sharing a common border have the same color. Use the minimum number of colors. You do not need to color the entire surface of each tile.

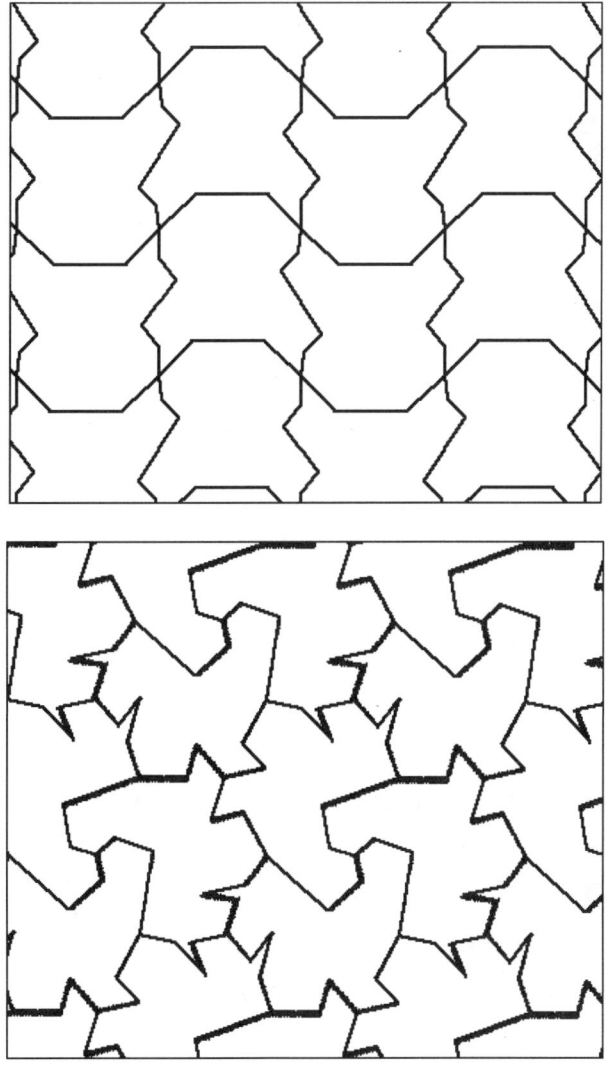

TesselMania! Math Connection

Minimal Coloring

Part 2

Use *TesselMania!* to find the minimum number of colors needed to color each of the tessellations listed below. Select each tile type, modify the tile as you wish, and use the paint bucket to fill it with color. Then create the tessellation. Record the number of colors that were needed.

Heesch Type	Minimum Number of Colors
Triangles	
CCC	
CGG	
Quadrilaterals	
TTTT	
TCTC	
GTGT	
$G_1 G_1 G_2 G_2$	
$G_1 G_2 G_1 G_2$	
GGCC	
GCGC	
CCCC	
$C_3 C_3 C_3 C_3$	
$C_3 C_3 C_6 C_6$	
$C_4 C_4 C_4 C_4$	
Hexagons	
TTTTTT	
$C_3 C_3 C_3 C_3 C_3 C_3$	

Summary

Describe your results.

_____

_____

_____

Conjecture

Make a conjecture regarding which characteristics of a tile or its tessellation are related to the minimum number of colors needed for the tessellation. Test your conjecture using *TesselMania!* Provide supporting evidence.

_____

_____

_____

Minimal Coloring
Activity 10: Teacher's Notes

Objectives

To determine the minimum number of colors needed to color tessellations
To relate the Heesch type of tessellation and the minimum number of colors
To make and test conjectures

Materials

Colored pencils, crayons, or markers
Student Activity Sheet 10

Prerequisites

Students must understand the Heesch system of classifying tessellations, as covered in the *TesselMania!* software and in Student Activities 4–7.

Notes

You may want to relate this activity to previous work students have done in map coloring. If you have advanced students who know that the minimum number of colors for map coloring is four, you may want to challenge them to explore whether any tessellations require four colors. (The authors are not aware of any.)

In Class

Have students work in pairs or small groups for this activity. Each student should color the two printed tessellations individually and compare results. (The coloring does not need to be complete or neat.) Be sure that students understand that tiles with a shared *border* must be different colors, but those with a shared *point (vertex)* can be the same color. Briefly discuss the results with the class.

Explain that students will use the *TesselMania!* software to examine each of the Heesch types and will determine the minimum number of colors needed for each type. Emphasize that they are collecting data that they will need to conjecture why some tessellations require two colors and others require three colors. The table helps them record and organize their data.

Encourage students to follow this procedure: choose each Heesch type, color the basic tile, tessellate it, look carefully at the tessellation for similarities and differences, and record the number of colors.

Students can share the work required to fill in the table and share their results with their partner(s). You may want to ask students to *predict* the minimum number of colors needed for each type of tessellation before they use the software to determine the actual number. They can write their predictions at the left of the column and their results at the right. Students can quickly see which Heesch types require two or three colors by noticing the color pairs or triplets offered in the color palette at the bottom of the main screen.

In the Summary section, encourage students to write their results and conclusions as clearly and as specifically as possible. In the Conjecture section, students may find it helpful to add a third column to the table so that they can record data such as the

Minimal Coloring
Teacher's Notes • page 2 of 3

number of touching vertices. Avoid telling students what to conjecture or how to conjecture. The process is more important than the product here.

Results

Part 1

The tessellation on the top (TCTC) requires two colors, and the tessellation on the bottom ($C_3C_3C_3C_3C_3C_3$) requires three colors.

Part 2

Heesch Type	Minimum Number of Colors	Number of Tiles Meeting at a Vertex
Triangles		
CCC	2	6
CGG	2	6
Quadrilaterals		
TTTT	2	4
TCTC	2	4
GTGT	2	4
$G_1G_1G_2G_2$	2	4
$G_1G_2G_1G_2$	2	4
GGCC	2	4
GCGC	2	4
CCCC	2	4
$C_3C_3C_3C_3$	3	3 or 6
$C_3C_3C_6C_6$	3	3, 4, or 6
$C_4C_4C_4C_4$	2	4
Hexagons		
TTTTTT	3	3
$C_3C_3C_3C_3C_3C_3$	3	3

Summary

Describe your results.

The smallest number of colors is two, and the largest number of colors is three. All the tessellations can be colored with either two or three colors. The tessellations near the bottom of this list require three colors.

Minimal Coloring
Teacher's Notes • page 3 of 3

Conjecture

Make a conjecture regarding which characteristics of a tile or its tessellation are related to the minimum number of colors needed for the tessellation. Test your conjecture using TesselMania! Provide supporting evidence.

Encourage students to test their conjectures with additional examples. This process may uncover counterexamples. Students should then revise their conjectures as necessary based on their testing, and write clear, precise conjecture statements.

The following conjectures were made by students during classroom testing. They have not been edited for content or style. They represent a good start, but they all need refinement. Some are valid and some are invalid. Many are not clear or specific enough to be readily understood by others.

> 1. All squares and equilateral triangle only need 2 colors; parallelograms and irregulars [sic] need 3 colors.
> 2. Angles over a certain measure need 3 colors.
> 3. If an angle of a figure is 120° or larger, the tessellation needs 3 colors.
> 4. Pick a point where the tiles meet. If the tile touches two others that touch each other, you need 3 colors.
> 5. Hexagons have 3 colors. Ones who touch two that touch each other need 3.
> 6. If there is only even numbers (of tiles) around one point, there are 2 colors. If there is an odd number (of tiles) around one point, there are 3 colors.
> 7. Ones with more or uneven sides were likely to require more colors.
> 8. The ones with either more than four sides or the ones with angles more than 90° need 3 colors.
> 9. Those above 90° have 3 colors minimum.

Here are some other possible conjectures. Some are valid, others are invalid.

Invalid
- If the basic tile has four sides, then two colors are needed; if it has three or six sides, then three colors are needed.
  Counterexample: TTTT needs two, but $C_3 C_3 C_3 C_3$ needs three.
- If rotations are involved, then three colors are needed; otherwise, two colors are needed.
  Counterexample: TTTTTT needs three, but CCC needs two.

Valid
- Three colors are needed if the original tile is a hexagon *or* if it is a quadrilateral *and* has a rotation of 120°.
- Three colors are needed if in the tessellation the number of tiles that meet at a vertex is three or six (or a multiple of three). Two colors are sufficient if the number of tiles at a vertex is four.
- Three colors are needed if exactly three tiles meet at least one vertex of the original tile.

TesselMania! Math Connection

Activity 11: Coordinate Connections

Explore how coordinate geometry makes *TesselMania!* possible.

Overview

How does *TesselMania!* create the transformed sides of tiles? How does it figure out how to move the tiles to tessellate? *You* can do this by sliding, turning, and flipping tracing paper, but a computer can't. Computer software needs an *algorithm*, which is a mathematical recipe. Coordinate geometry comes to the rescue!

Coordinate geometry makes it possible to draw tiles and tessellations on the computer screen. Each *pixel* (point) on the computer screen corresponds to a point on a coordinate grid. A point A can be represented by its coordinates (x, y).

The *image* of point A after a transformation has been performed is labeled A' and has the coordinates (x', y'). The original point A is called the *pre-image*.

In this activity you explore ways to describe transformations, using coordinate points and simple equations. In Part A you work with translations. Parts B and C focus on glide reflections and rotations.

Part A. Translations

1. Translating a Point

In the figure below, point A is translated to form the image point A'. Write two equations that give the coordinates of image point A' if you know the coordinates of point A. The equation for x' is given as an example.

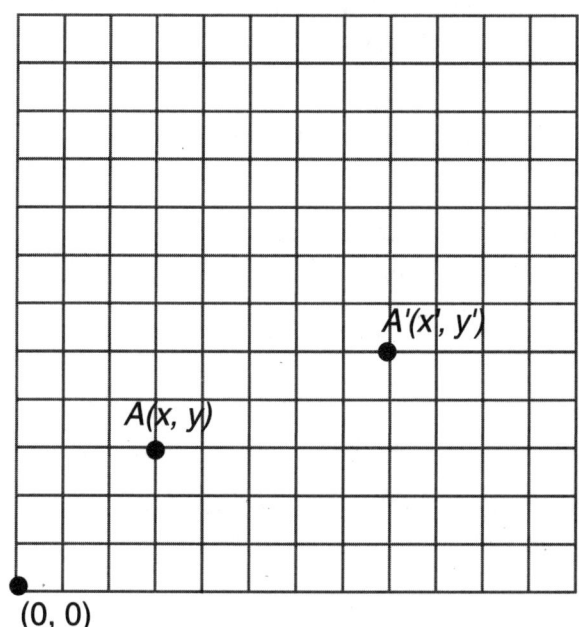

The amount of movement in the x direction from A to A' is +5. This is called the *displacement*, or Δx. (Sometimes Δx is called the *change in x*.)

$$\Delta x = +5$$

The value of x' depends on the original x value and the displacement Δx. The equation for x' is given below.

$$x' = x + 5$$

What is the displacement of y in this figure?

$\Delta y = $ _____

The value of y' depends on the original y value and the displacement Δy.

$y' = $ _____

Write two *general* equations for x' and y' when point A is translated to A'. (Hint: Use x and y, and use Δx and Δy in place of numbers.)

$x' = $ _____ $y' = $ _____

Coordinate Connections

page 2 of 9

2. Translating a Tile Side

In the tile below, one side is missing. Draw the missing side formed by translation.
- List the coordinates of the marked points.
- Find the Δx and Δy for this translation. $\Delta x =$ _____ $\Delta y =$ _____
- Write equations for x' and y'. Use them to calculate the coordinates (x', y') of the image points. $x' =$ _____ $y' =$ _____
- Plot the image points on the grid. Connect them with line segments.

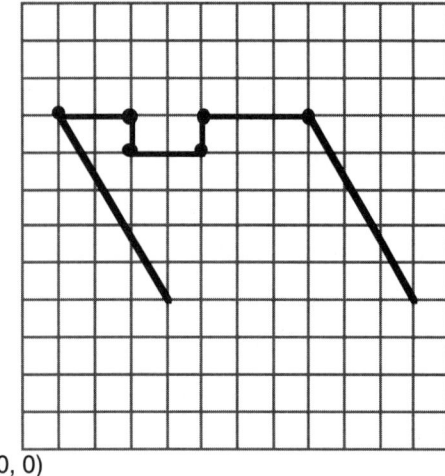

(0, 0)

Pre-Image (x, y)	Image (x', y')
(1, 9)	(__, __)
(__, __)	(__, __)
(__, __)	(__, __)
(__, __)	(__, __)
(__, __)	(__, __)
(__, __)	(__, __)

3. Translating Tiles

Recall how a translated tile (TTTT) is tessellated (or use the animated tessellation button on *TesselMania!* to see it again). When a quadrilateral tile is formed by translation, the eight adjacent tiles are formed by translating the original tile.

In the figure below, the coordinates of the six vertices of the motif tile are shown. Calculate the coordinates for the three adjacent tiles. List the Δx and Δy values for each translation. Are they the same or different?

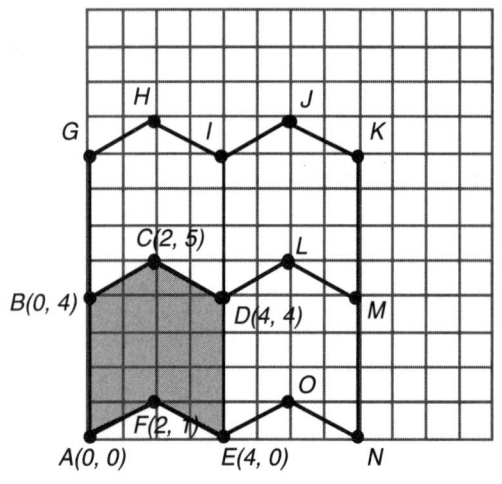

Motif Tile	Top	Diagonal	Side
A (0, 0)	B (0, 4)	D (4, 4)	E (4, 0)
B (0, 4)	G (__, __)	I (__, __)	D (4, 4)
C (2, 5)	H (__, __)	J (__, __)	L (__, __)
D (4, 4)	I (__, __)	K (__, __)	M (__, __)
E (4, 0)	D (4, 4)	M (__, __)	N (__, __)
F (2, 1)	C (2, 5)	L (__, __)	O (__, __)
	$\Delta x =$	$\Delta x =$	$\Delta x =$
	$\Delta y =$	$\Delta y =$	$\Delta y =$

Coordinate Connections

4. Matrices and Translations

Calculating each image point one by one for the vertices of a complicated polygon is tedious work. Computer software employs a streamlined method using *matrices*. To represent translations, use the addition of matrices.

The point $A(x, y)$ can be represented by a matrix: $\begin{bmatrix} x \\ y \end{bmatrix}$. The amount of the translation can also be represented by a matrix: $\begin{bmatrix} \Delta x \\ \Delta y \end{bmatrix}$. The coordinates of image point A' can be calculated by adding these matrices, as shown below.

$$\begin{bmatrix} \Delta x \\ \Delta y \end{bmatrix} + \begin{bmatrix} x \\ y \end{bmatrix} = \begin{bmatrix} \Delta x + x \\ \Delta y + y \end{bmatrix} = \begin{bmatrix} x' \\ y' \end{bmatrix}$$

For example, in the tessellation in Step 3, the point $C(2, 5)$ translates to point $J(6, 9)$;

$$\begin{bmatrix} 4 \\ 4 \end{bmatrix} + \begin{bmatrix} 2 \\ 5 \end{bmatrix} = \begin{bmatrix} 4+2 \\ 4+5 \end{bmatrix} = \begin{bmatrix} 6 \\ 9 \end{bmatrix}$$ where $\Delta x = 4$ and $\Delta y = 4$.

Show the matrix calculations that find the other points of this tile: $D, I, (J), K, M, L$.

Pre-Image Point	Image Point	Matrix Calculations
	D	
	I	
C	J	$\begin{bmatrix} 4 \\ 4 \end{bmatrix} + \begin{bmatrix} 2 \\ 5 \end{bmatrix} = \begin{bmatrix} 6 \\ 9 \end{bmatrix}$
	K	
	M	
	L	

5. Summary

Use coordinate geometry concepts in your summaries.

1. Write a brief description of what the computer software does to calculate an image point when you "bump" the top of a square tile up to a new point (type TTTT).

2. Write a brief description of how the computer software calculates where to draw the edges of the eight tiles adjacent to the motif.

Coordinate Connections

Part B. Glide Reflections

This activity explores glide reflections in two special cases: reflecting across a vertical line and across a horizontal line. Recall that a glide reflection is a reflection and a translation combined, but the translation must be parallel to the line of reflection.

1. Glide-Reflecting a Tile Side

Glide-reflect this tile side across the y-axis. Use $\Delta y = 4$ ($\Delta x = 0$) for the translation.

- Draw the image points after the reflection and then after the translation.
- Write the coordinates of pre-image points and image points in the columns below.
- Write equations for x' and y' that represent how to find the image coordinates when you have the pre-image coordinates (x, y), the value of Δy, and a reflection across the y-axis.

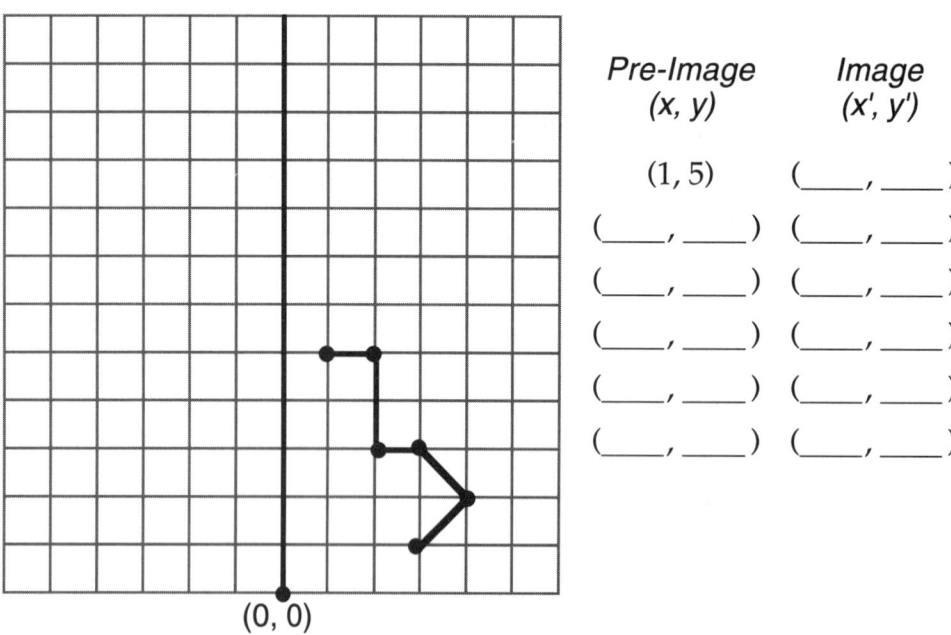

Pre-Image (x, y)	Image (x', y')
(1, 5)	(__, __)
(__, __)	(__, __)
(__, __)	(__, __)
(__, __)	(__, __)
(__, __)	(__, __)
(__, __)	(__, __)

Glide Reflection Across the Y-axis

$x' = $ _____

$y' = $ _____

Glide Reflection Across a Vertical Line

If the vertical line of reflection is *not* the y-axis ($x = 0$), but the line $x = n$ where n is some number, then use the equations below.

$x' = -x + 2n$
$y' = y + \Delta y$

Use these equations to find the image of point (5, 8) when it is reflected across the line $x = 2$ and $\Delta y = 3$. Plot and label the pre-image point (5, 8) and its image on the grid above.

Coordinate Connections

page 5 of 9

Glide-reflect this next tile side across the *x*-axis. Use $\Delta x = -2$ ($\Delta y = 0$) for the translation.

- Draw the image points after the reflection and then after the translation.
- Complete the columns of pre-image points and image points by listing coordinates.
- Write equations for x' and y' that represent how to find the image coordinates when you have the pre-image coordinates (x, y), the value of Δy, and a reflection across the *x*-axis.

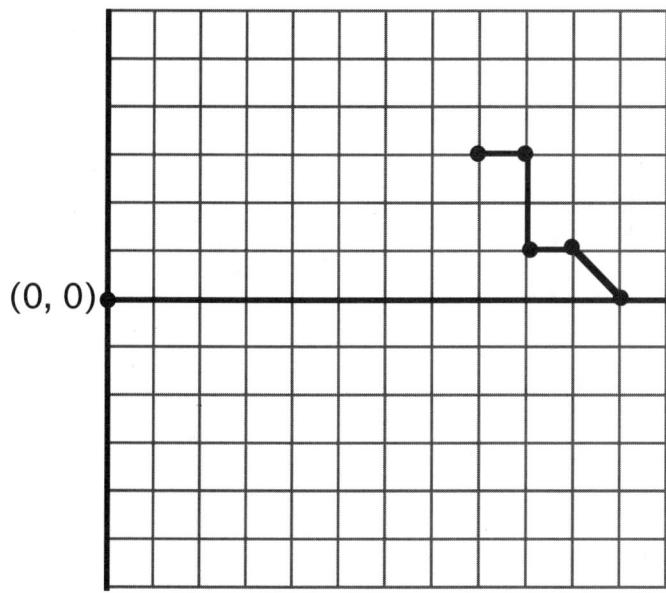

Pre-Image (x, y)	Image (x', y')
(8, 3)	(___, ___)
(___, ___)	(___, ___)
(___, ___)	(___, ___)
(___, ___)	(___, ___)
(___, ___)	(___, ___)
(___, ___)	(___, ___)

Glide Reflection Across the *X*-axis

$x' =$ _____

$y' =$ _____

Glide Reflection Across a Horizontal Line

If the horizontal line of reflection is *not* the *x*-axis ($y = 0$), but the line $y = n$, then use the equations below.

$x' = x + \Delta x$
$y' = -y + 2n$

Use these equations to find the image of point (4, 1) when it is reflected across the line $y = 3$ and $\Delta x = 3$. Plot and label the pre-image point (4, 1) and its image on the grid above.

Challenge

Does it make a difference to the final image point under a glide reflection whether the reflection or the translation is done first? Show the evidence you find to support your answer.

Coordinate Connections

page 6 of 9

2. Glide-Reflecting Tiles

Recall how a $G_1 G_2 G_1 G_2$ tile is tessellated (or use the animation tessellation button on *TesselMania!* to see it again). The four adjacent tiles at the top, bottom, right side, and left side are formed by glide reflections. The diagonal tiles are formed by rotations of 180°. (Rotations in a tessellation formed by $G_1 G_2 G_1 G_2$ seem mysterious. You can investigate this further in the Challenge activity.)

In the figure below, find the coordinates of the top and side adjacent tiles. You can use the equations from Part 1, or you can read them off the grid. Do you get the same results either way?

For the top tile, the line of reflection is $x = 2$ and the translation is $\Delta y = 4$. For the side tile, the line of reflection is $y = 2$ and the translation is $\Delta x = 4$.

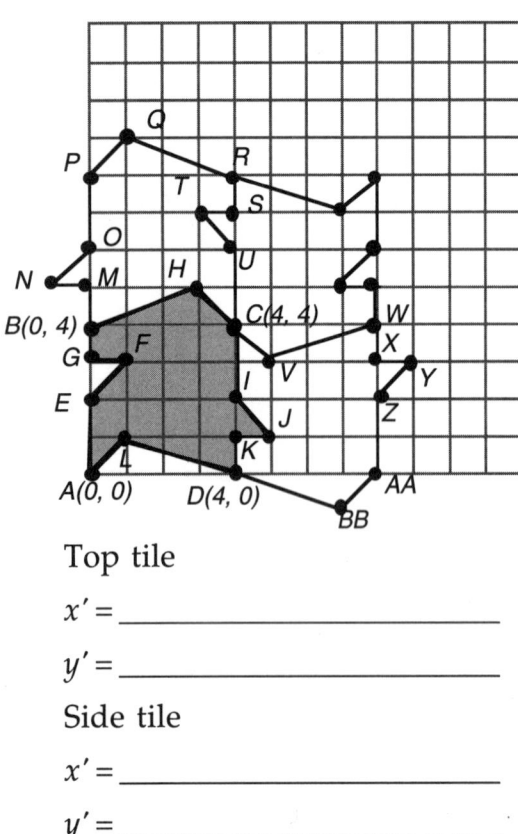

Motif Tile		Top		Side	
A	(0, 0)	C	(4, 4)	C	(4, 4)
B	(0, 4)	R	(__, __)	D	(4, 0)
C	(4, 4)	P	(__, __)	AA	(__, __)
D	(4, 0)	B	(0, 4)	W	(__, __)
E	(0, 2)	U	(__, __)	I	(4, 2)
F	(1, 3)	T	(__, __)	J	(5, 1)
G	(0, 3)	S	(__, __)	K	(4, 1)
H	(3, 5)	Q	(__, __)	BB	(__, __)
I	(4, 2)	O	(__, __)	Z	(__, __)
J	(5, 1)	N	(__, __)	Y	(__, __)
K	(4, 1)	M	(__, __)	X	(__, __)
L	(1, 1)	H	(3, 5)	V	(__, __)

Top tile

$x' = $ _____

$y' = $ _____

Side tile

$x' = $ _____

$y' = $ _____

You can see why a computer is useful here! It can do these calculations very quickly.

Challenge

Experiment to see if there is a way to place the diagonal adjacent tile using glide reflections instead of a rotation of 180°. (Hint: The top tile is placed by a glide reflection across a vertical line; the side tile reflects across a horizontal line.) Show the calculations and drawings you used to reach your conclusion. Write a brief description of your method.

Coordinate Connections

3. Matrices and Glide Reflections

You already know how to use a matrix to represent a translation. In this activity you explore how to use a matrix to represent a reflection across the *x*- or *y*-axis. Then you combine these two procedure to represent a glide reflection.

Matrix multiplication, along with a special matrix, can represent reflection across the *y*-axis. To multiply two matrices, the number of columns of the first matrix must equal the number of rows of the second matrix. Since a point is represented by a matrix with two rows, the special matrix must have two columns.

The special matrix for reflection across the *y*-axis is $\begin{bmatrix} -1 & 0 \\ 0 & 1 \end{bmatrix}$.

If the pre-image point *A* is (3, 2), image *A'* after reflection across the *y*-axis is (–3, 2).

$$\begin{bmatrix} -1 & 0 \\ 0 & 1 \end{bmatrix} \begin{bmatrix} 3 \\ 2 \end{bmatrix} = \begin{bmatrix} -1 \times 3 + 0 \times 2 \\ 0 \times 3 + 1 \times 2 \end{bmatrix} = \begin{bmatrix} -3 \\ 2 \end{bmatrix}$$

Find a different special matrix that represents reflection across the *x*-axis. Write it below.

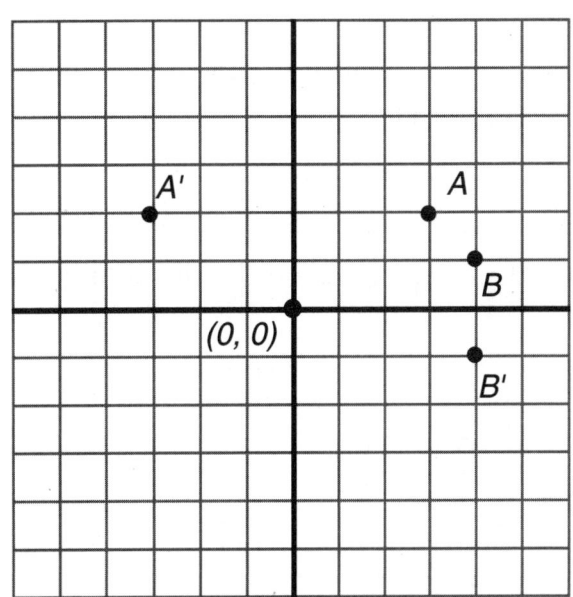

Find a two-step process that uses matrices to calculate the coordinates of an image point under a glide reflection across the *y*-axis with a translation of $\Delta y = 4$. Let the pre-image point be (*x*, *y*). (Choose a specific point, like (6, 3), to test your process.)

Find a two-step process that uses matrices to calculate the coordinates of an image point under a glide reflection across the *x*-axis with a translation of $\Delta x = 6$. Let the pre-image point be (*x*, *y*). (Choose a specific point to test your process.)

Coordinate Connections

Part C. Rotations

Finding equations for image point coordinates under rotation is somewhat complicated. The equations use trigonometric functions, like sine and cosine. This activity uses only rotations of 90°, and the center of rotation is always the origin, so the equations will be simpler.

1. Rotating a Tile Side

If α is the angle of rotation, then equations for the image points under rotation are as shown below.

$x' = x \cos \alpha - y \sin \alpha$

$y' = x \sin \alpha + y \cos \alpha$

Let α be 90°. You may know that sin 90° = 1, cos 90° = 0.

Rewrite the equations with these values for sin and cos.

Rotation of 90°

$x' = $ _____

$y' = $ _____

Use these equations on the tile side below.

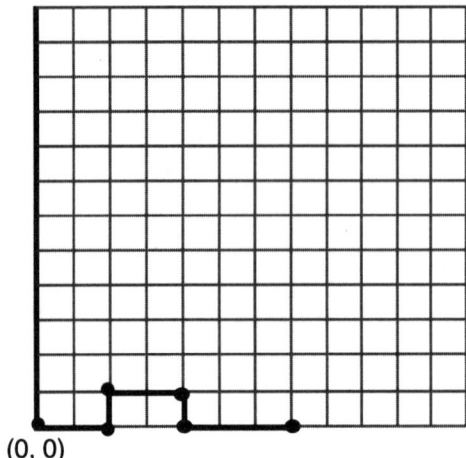

(0, 0)

Pre-Image (x, y)	Image (x', y')
(0, 0)	(___, ___)
(___, ___)	(___, ___)
(___, ___)	(___, ___)
(___, ___)	(___, ___)
(___, ___)	(___, ___)
(___, ___)	(___, ___)

Coordinate Connections

2. Rotating Tiles

Recall how the tile type $C_4 C_4 C_4 C_4$ is tessellated (or review it using *TesselMania!*). The top, bottom, left side, and right side adjacent tiles are rotated 90°. The diagonal adjacent tiles are rotated 180°. Calculate the coordinate of the tile on the left side. It is rotated 90° around point *A*.

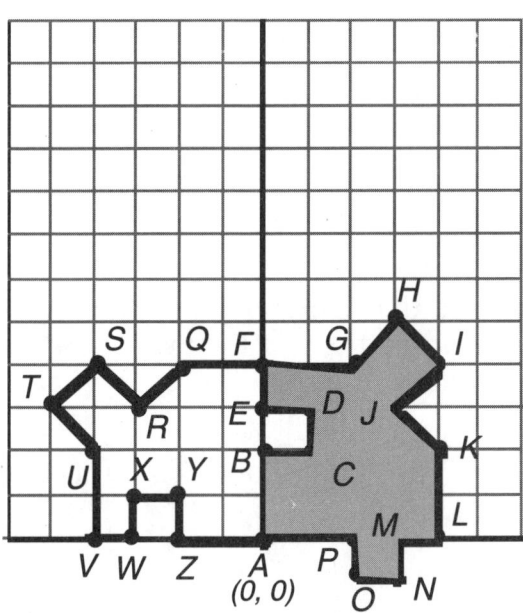

Motif Tile		Left Side	
A	(0, 0)	A	(0, 0)
B	(0, 2)	Z	(___, ___)
C	(1, 2)	Y	(___, ___)
D	(1, 3)	X	(___, ___)
E	(0, 3)	W	(___, ___)
F	(0, 4)	V	(___, ___)
G	(2, 4)	U	(___, ___)
H	(3, 5)	T	(___, ___)
I	(4, 4)	S	(___, ___)
J	(3, 3)	R	(___, ___)
K	(4, 2)	Q	(___, ___)
L	(4, 0)	F	(0, 4)
M	(3, 0)	E	(0, 3)
N	(3, −1)	D	(1, 3)
O	(2, −1)	C	(1, 2)
P	(2, 0)	B	(0, 2)

Challenge

Write equations for the image point coordinates when the center of rotation (90°) is not the origin. For example, what if the center of rotation is point *I* in the figure above?

Coordinate Connections

Activity 11: Teacher's Notes

Objective

To explore how coordinate geometry makes *TesselMania!* possible
To explore translations and special glide reflections and rotations using coordinates
To write equations to describe image points under transformations
To use matrices to represent transformations

Materials

Student Activity Sheet 11
Transparency Master: Activity 11

Prerequisites

Students should have completed Activities 4 and 5 so that they understand how tiles are formed by transforming the sides and how transformations of the motif create a tessellation. They should be able to write the coordinates for a point on a labeled coordinate grid and be able to use variables to create equations.

In Part C, students need to understand and be able to use the trigonometric functions sine and cosine. Parts A and B do not require trigonometry.

Background

This activity consists of three parts: translation, glide reflection, and rotation. Each part has an activity based on creating a tile, another on creating a tessellation, and a third on matrix representation.

You will want to adjust this activity, depending on the amount of experience your students have had with coordinate geometry. If this topic is relatively new to them, you may want to use only Part A. Do Steps 1 and 2 together with the whole class, then have students do Step 3 and the summary in Step 5 on their own. You may want to omit Step 4, which involves matrices.

If your students are experienced in using coordinate geometry, you may want to have them work independently on these activity sheets. In some cases, you may choose to do Part A quickly and spend more time on the more advanced Parts B and C.

In Class

Pose the question: How can the *TesselMania!* software "know" where to draw the transformed side and the tessellated tiles? Ask students to recall how they created tiles and tessellations by hand with tracing paper. Point out that computers can't turn and flip paper or the screen. Computers need an *algorithm*, or set of instructions that can be thought of as a mathematical recipe.

Coordinate geometry is a tool we can use to create an algorithm. We place the tile on a coordinate grid, and each vertex is represented by a coordinate point (x, y). Each point on the grid corresponds to a *pixel* on the computer screen. A *pixel* (short for *picture element*) is a tiny dot of light or dark on the screen. (Of course, a mathematical point has a diameter of zero, while a pixel has a non-zero diameter. Students who

Coordinate Connections

Teacher's Notes

have studied programming may know that the coordinate grid system on the computer screen does not exactly match the one used in mathematics, where (0, 0) is at the lower left, and *y*-values increase from the bottom to the top. However, it is not necessary to emphasize these issues here.)

Part A. Translations

1. **Translating a Point**

    Explain that the whole class will do an example of translation using coordinate geometry. (You may choose to do the example that is worked on the student activity sheet or a similar example with different points and translation values.)

    On an overhead or the board, label point $A(x, y)$. If necessary, consider specific examples, such as (2, 6) or (5, 3). Then suppose we translate this point A right five units and up two units. Place and label the image point A' (read "A-prime"). Be sure students are clear on the use of the terms *image* and *pre-image*.

    Since we aren't yet sure what the coordinates of A' are, we will label them (x', y'). We can find the actual coordinates of A'. Ask students how to do this (add five to the *x*-value, add two to the *y*-value). Point out that the change in the *x*-value is +5; this can be called Δx, or *displacement of x*. Ask students the value of Δy (Answer: 2).

    Ask them how to write an equation that expresses x' in terms of x and Δx (Answer: $x' = x + 5$). Similarly, write an equation for y' (Answer: $y' = y + 2$). Write both equations on the overhead or the board.

    Ask how to write a general equation using Δx and Δy instead of their specific values. Write these equations also: $(x' = x + \Delta x, y' = y + \Delta y)$. Have students record these results in Step 1 on their activity sheets.

2. **Translating a Tile Side**

    Direct students to continue with Step 2, in which they calculate the image points for the bottom side of a tile. Briefly discuss their results as a whole class.

3. **Translating Tiles**

    Introduce the question: How can coordinate geometry determine where to draw the vertices of the tessellated tiles that are adjacent to the motif (the original tile)?

    If needed, remind students of how the TTTT tile tessellates, using the middle "magic" button to see the animation on *TesselMania!* Ask students for ideas on how to calculate the image of the motif for each of the eight adjacent tiles. (Each motif vertex point is translated to its image point; the translation—Δx and Δy values—is different for each of the eight adjacent tiles.)

    You may want to have students complete this activity with their partners, or you can do it together as a whole class.

Coordinate Connections
Teacher's Notes • page 3 of 8

4. **Matrices and Translations**

   Explain that this activity introduces the use of a matrix to represent both a coordinate point and the amount of a translation. Matrix addition is used to calculate the image point coordinates under translation.

   Follow the student activity sheet to work through the example of point C and image point J in the tile positioned diagonally from the motif. Direct students to follow the same process for the other five points of this tile and record their results on the table.

5. **Summary**

   Direct students to write the two descriptions in the Summary section. Have students share their descriptions in small groups or with the whole class.

Results

Part A. Translations

1. **Translating a Point**

Displacement for x and for y	$\Delta x = 5$	$\Delta y = 2$
Equations for x' and y'	$x' = x + 5$	$y' = y + 2$
General equations	$x' = x + \Delta x$	$y' = y + \Delta y$

2. **Translating a Tile Side**

   $\Delta x = 3 \qquad \Delta y = -5$

   $x' = x + 3 \qquad y' = y - 5$

Pre-Image	Image
(1, 9)	(4, 4)
(3, 9)	(6, 4)
(3, 8)	(6, 3)
(5, 8)	(8, 3)
(5, 9)	(8, 4)
(8, 9)	(11, 4)

Coordinate Connections
Teacher's Notes • page 4 of 8

3. Translating Tiles

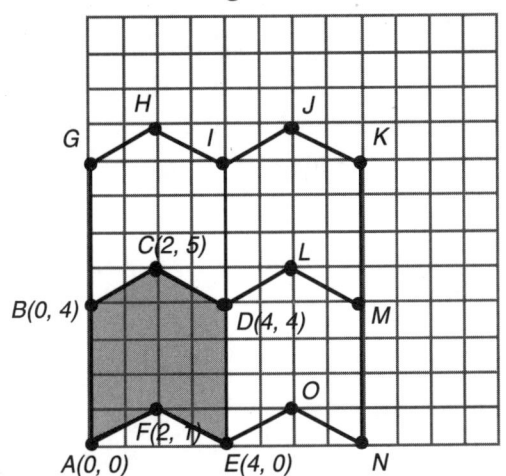

Motif Tile	Top	Diagonal	Side
A (0, 0)	B (0, 4)	D (4, 4)	E (4, 0)
B (0, 4)	G (0, 8)	I (4, 8)	D (4, 4)
C (2, 5)	H (2, 9)	J (6, 9)	L (6, 5)
D (4, 4)	I (4, 8)	K (8, 8)	M (8, 4)
E (4, 0)	D (4, 4)	M (8, 4)	N (8, 0)
F (2, 1)	C (2, 5)	L (6, 5)	O (6, 1)
	Δx = 0	Δx = 4	Δx = 4
	Δy = 4	Δy = 4	Δy = 0

4. Matrices and Translations

Pre-Image Point	Image Point	Matrix Calculations
A	D	$\begin{bmatrix}4\\4\end{bmatrix}+\begin{bmatrix}0\\0\end{bmatrix}=\begin{bmatrix}4\\4\end{bmatrix}$
B	I	$\begin{bmatrix}4\\4\end{bmatrix}+\begin{bmatrix}0\\4\end{bmatrix}=\begin{bmatrix}4\\8\end{bmatrix}$
C	J	$\begin{bmatrix}4\\4\end{bmatrix}+\begin{bmatrix}2\\5\end{bmatrix}=\begin{bmatrix}6\\9\end{bmatrix}$
D	K	$\begin{bmatrix}4\\4\end{bmatrix}+\begin{bmatrix}4\\4\end{bmatrix}=\begin{bmatrix}8\\8\end{bmatrix}$
E	M	$\begin{bmatrix}4\\4\end{bmatrix}+\begin{bmatrix}4\\0\end{bmatrix}=\begin{bmatrix}8\\4\end{bmatrix}$
F	L	$\begin{bmatrix}4\\4\end{bmatrix}+\begin{bmatrix}2\\1\end{bmatrix}=\begin{bmatrix}6\\5\end{bmatrix}$

5. Summary

1. Write a brief description of what the computer software does to calculate an image point when you "bump" the top of a square tile up to a new point (type TTTT).

   When you "bump" the top side to a new point, the computer software "knows" the coordinates of this new point. It calculates the coordinates of the image point, using equations or matrices for translation, and plots that image point. Then it draws line segments from the bottom vertices to the image point.

Coordinate Connections
Teacher's Notes • page 5 of 8

2. Write a brief description of how the computer software calculates where to draw the edges of the eight tiles adjacent to the motif.

   The software "knows" the coordinates of each vertex of the motif tile. It calculates the image points, using the equations or matrices for translations. Each of the eight adjacent tiles uses a different pair of Δx and Δy values for translation. The image points are plotted, and line segments are drawn between them to create the adjacent tile.

Part B. Glide Reflection

1. Glide Reflecting a Tile Side

Glide Reflect Across the Y-axis

Reflect across y-axis; translate $\Delta y = 4$.

Pre-Image (x, y)	Image (x', y')
(1, 5)	(−1, 9)
(2, 5)	(−2, 9)
(2, 3)	(−2, 7)
(3, 3)	(−3, 7)
(4, 2)	(−4, 6)
(3, 1)	(−3, 5)

$x' = -x$
$y' = y + \Delta y$

Glide Reflect Across the X-axis

Reflect across x-axis; translate $\Delta x = -2$.

Pre-Image (x, y)	Image (x', y')
(8, 3)	(6, −3)
(9, 3)	(7, −3)
(9, 1)	(7, −1)
(10, 1)	(8, −1)
(11, 0)	(9, 0)

$x' = x + \Delta x$
$y' = -y$

Glide Reflect Across a Vertical Line

Find the image points of point (5, 8) when it is reflected across the line $x = 2$ and $\Delta y = 3$.

(−1, 11)

Glide Reflect Across a Horizontal Line

Find the image points of point (4, 1) when it is reflected across the line $y = 3$ and $\Delta x = 3$.

(7, 5)

Challenge

Does it make a difference to the final image point under a glide reflection whether the reflection or the translation is done first? Show the evidence you find to support your answer.

No, it does not make any difference in the image point whether the reflection or the translation is done first. Evidence might include worked examples showing the intermediate image points and the final image point under both methods.

Coordinate Connections
Teacher's Notes • page 6 of 8

2. Glide-Reflecting Tiles

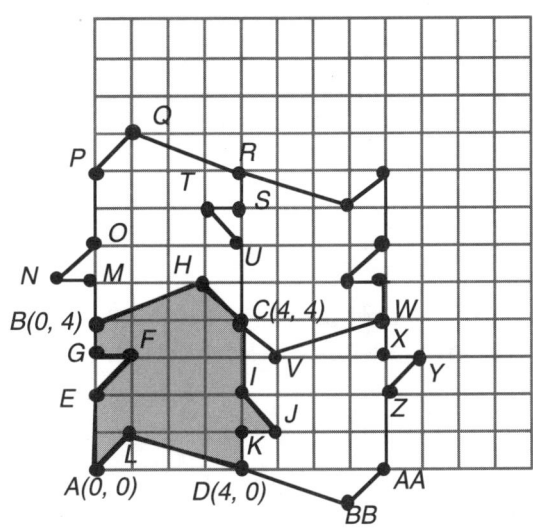

Motif Tile	Top	Side
A (0, 0)	C (4, 4)	C (4, 4)
B (0, 4)	R (4, 8)	D (4, 0)
C (4, 4)	P (0, 8)	AA (8, 0)
D (4, 0)	B (0, 4)	W (8, 4)
E (0, 2)	U (4, 6)	I (4, 2)
F (1, 3)	T (3, 7)	J (5, 1)
G (0, 3)	S (4, 7)	K (4, 1)
H (3, 5)	Q (1, 9)	BB (7, −1)
I (4, 2)	O (0, 6)	Z (8, 2)
J (5, 1)	N (−1, 5)	Y (9, 3)
K (4, 1)	M (0, 5)	X (8, 3)
L (1, 1)	H (3, 5)	V (5, 3)

Top tile
$x' = -x + 4$
$y' = y + 4$

Side tile
$x' = x + 4$
$y' = -y + 4$

Challenge

Experiment to see if there is a way to place the diagonal adjacent tile using glide reflections instead of a rotation of 180°. (Hint: The top tile is placed by a glide reflection across a vertical line; the side tile reflects across a horizontal line.) Show the calculations and drawings you used to reach your conclusion. Write a brief description of your method.

The composition of two glide reflections is a rotation. Combining a glide reflection across a vertical line and a glide reflection across a horizontal line results in the same image as a rotation of 180°. In this example, the diagonal tile could be placed by reflecting the motif across the vertical line $x = 4$, followed by a reflection across the horizontal line $y = 4$.

3. Matrices and Glide Reflection

Reflection across the y-axis $\begin{bmatrix} -1 & 0 \\ 0 & 1 \end{bmatrix}$.

Reflection across the x-axis $\begin{bmatrix} 1 & 0 \\ 0 & -1 \end{bmatrix}$

Coordinate Connections
Teacher's Notes • page 7 of 8

Find a two-step process that uses matrices to calculate the coordinates of an image point under a glide reflection across the y-axis with a translation of $\Delta y = 4$.

The process involves multiplying by the reflection matrix, then adding the translation matrix.

First glide-reflect across the y-axis.
$$\begin{bmatrix} -1 & 0 \\ 0 & 1 \end{bmatrix} \begin{bmatrix} x \\ y \end{bmatrix} = \begin{bmatrix} -x \\ y \end{bmatrix}$$

Then translate with $\Delta y = 4$.
$$\begin{bmatrix} 0 \\ 4 \end{bmatrix} + \begin{bmatrix} -x \\ y \end{bmatrix} = \begin{bmatrix} -x \\ y+4 \end{bmatrix}$$

Find a two-step process that uses matrices to calculate the coordinates of an image point under a glide reflection across the x-axis with a translation of $\Delta x = 6$.

First glide-reflect across the x-axis.
$$\begin{bmatrix} 1 & 0 \\ 0 & -1 \end{bmatrix} \begin{bmatrix} x \\ y \end{bmatrix} = \begin{bmatrix} x \\ -y \end{bmatrix}$$

Then translate with $\Delta x = 6$.
$$\begin{bmatrix} 6 \\ 0 \end{bmatrix} + \begin{bmatrix} x \\ -y \end{bmatrix} = \begin{bmatrix} 6+x \\ -y \end{bmatrix}$$

Part C. Rotations

1. Rotating a Tile Side

Rotation of 90°

$x' = -y \qquad y' = x$

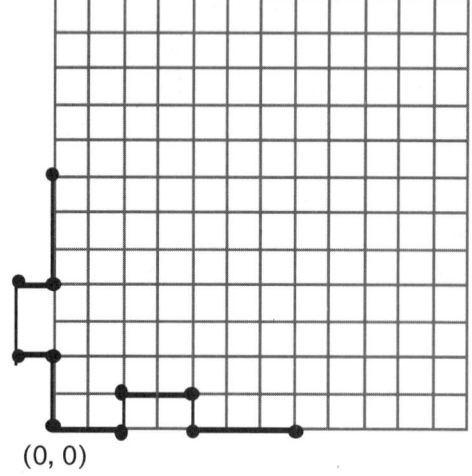

(0, 0)

Pre-Image (x, y)	Image (x', y')
(0, 0)	(0, 0)
(2, 0)	(0, 2)
(2, 1)	(−1, 2)
(4, 1)	(−1, 4)
(4, 0)	(0, 4)
(7, 0)	(0, 7)

Coordinate Connections
Teacher's Notes • page 8 of 8

2. Rotating Tiles

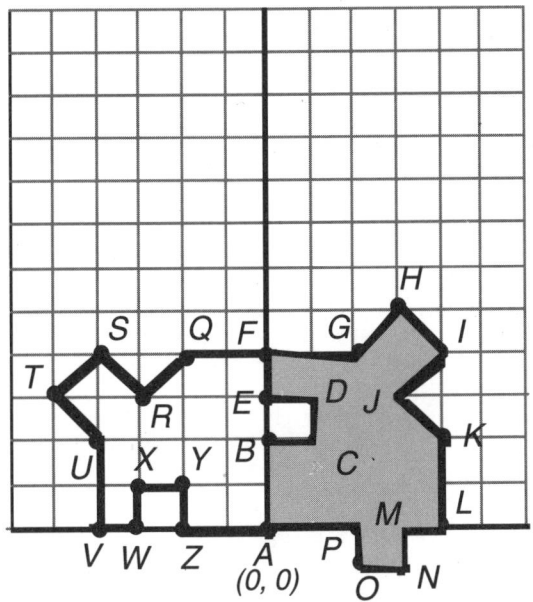

Motif Tile		Left Side	
A	(0, 0)	A	(0, 0)
B	(0, 2)	Z	(−2, 0)
C	(1, 2)	Y	(−2, 1)
D	(1, 3)	X	(−3, 1)
E	(0, 3)	W	(−3, 0)
F	(0, 4)	V	(−4, 0)
G	(2, 4)	U	(−4, 2)
H	(3, 5)	T	(−5, 3)
I	(4, 4)	S	(−4, 4)
J	(3, 3)	R	(−3, 3)
K	(4, 2)	Q	(−2, 4)
L	(4, 0)	F	(0, 4)
M	(3, 0)	E	(0, 3)
N	(3, −1)	D	(1, 3)
O	(2, −1)	C	(1, 2)
P	(2, 0)	B	(0, 2)

Challenge

Write equations for the image point coordinates when the center of rotation (90°) is not the origin. For example, what if the center of rotation is point I in the figure above?

$x' = -y + 2n$, $y' = x$ where $n = 4$ because I is (4, 4).

Motif Tile		Image		Motif Tile		Image	
A	(0, 0)		**(8, 0)**	G	(2, 4)	K	**(4, 2)**
B	(0, 2)		**(6, 0)**	H	(3, 5)	J	**(3, 3)**
C	(1, 2)		**(6, 1)**	I	(4, 4)	I	**(4, 4)**
D	(1, 3)		**(5, 1)**	J	(3, 3)		**(5, 3)**
E	(0, 3)		**(5, 0)**	K	(4, 2)		**(6, 4)**
F	(0, 4)	L	**(4, 0)**	L	(4, 0)		**(8, 4)**

Transparency Master
Activity 1

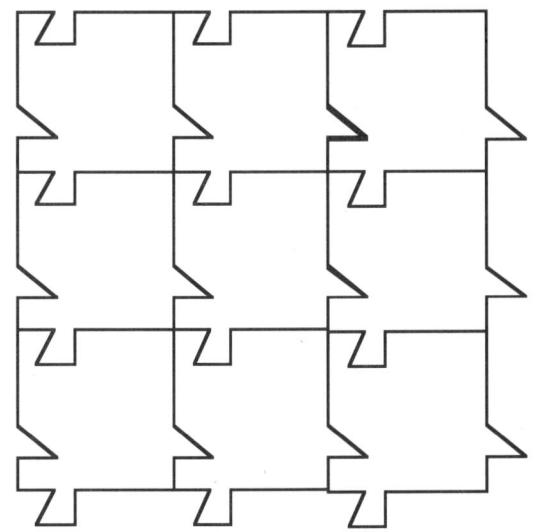
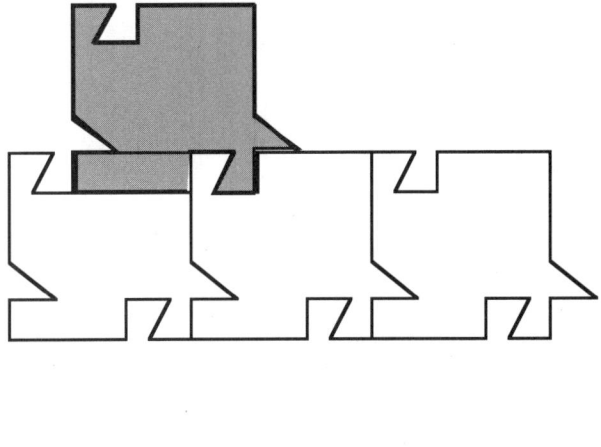
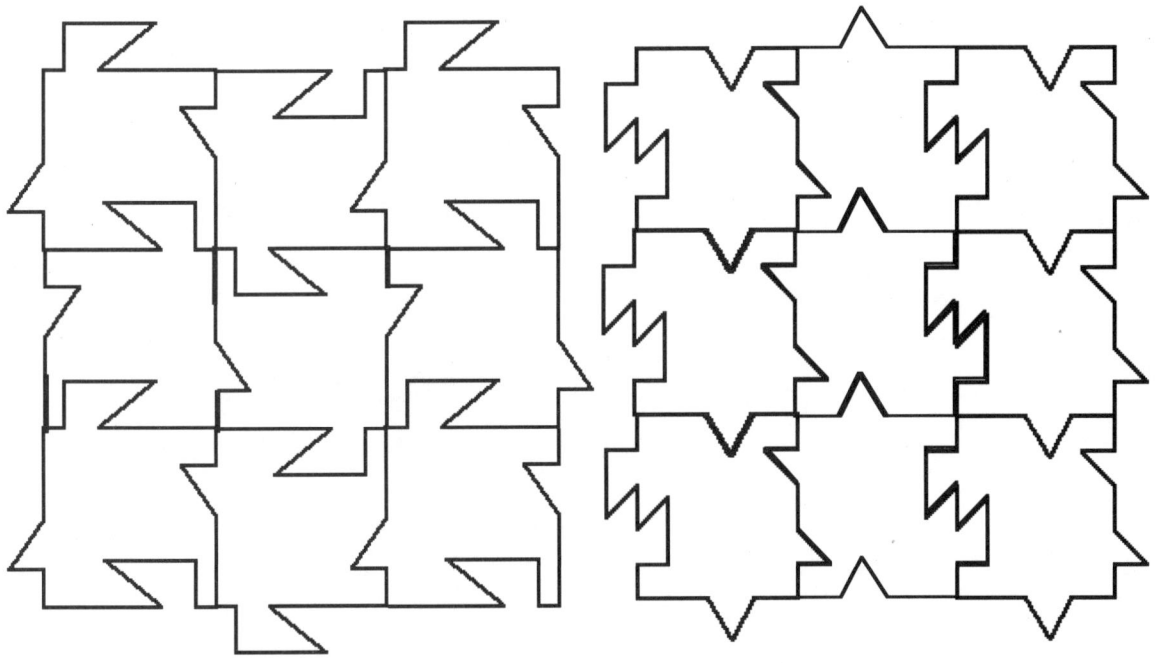

TesselMania! Math Connection

©1995 by Key Curriculum Press
Do not copy without permission.

Transparency Master • 87

Transparency Master
Activity 4

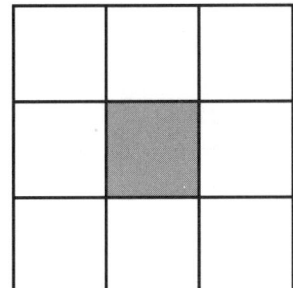   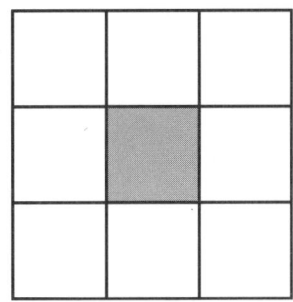

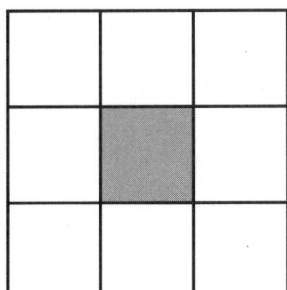   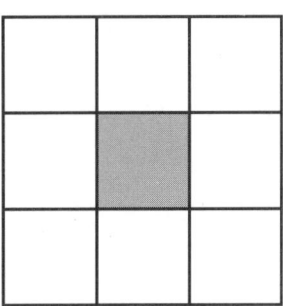

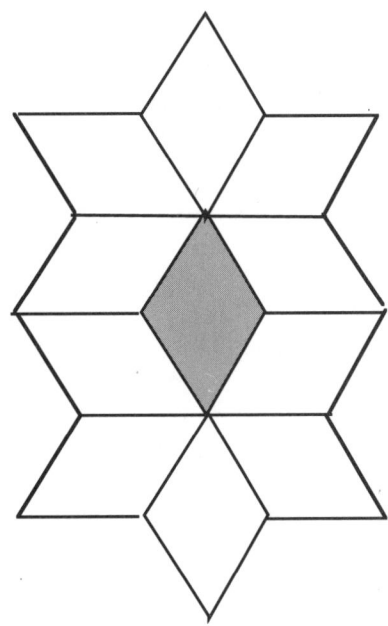

TesselMania! Math Connection

Transparency Master • 89

Transparency Master
Activities 5 and 6

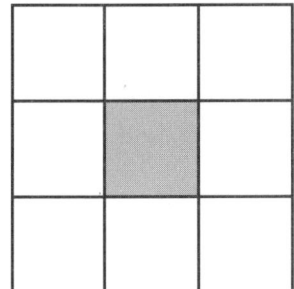 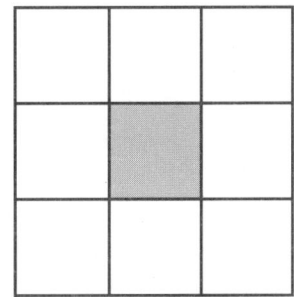

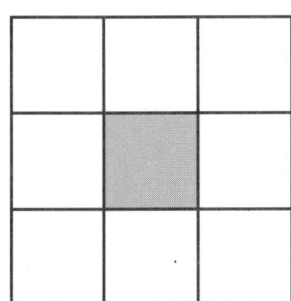 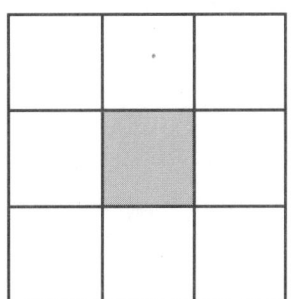

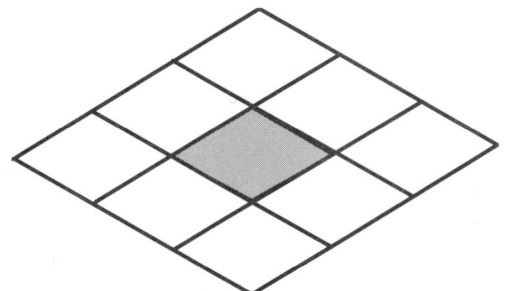 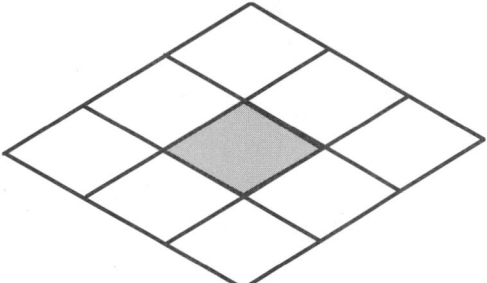

TesselMania! Math Connection

Transparency Master
Activity 7

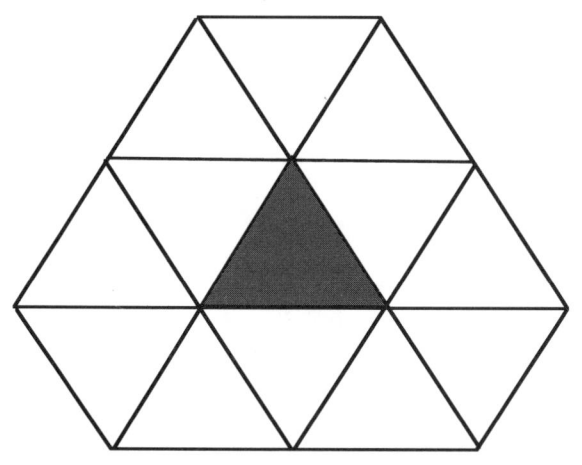
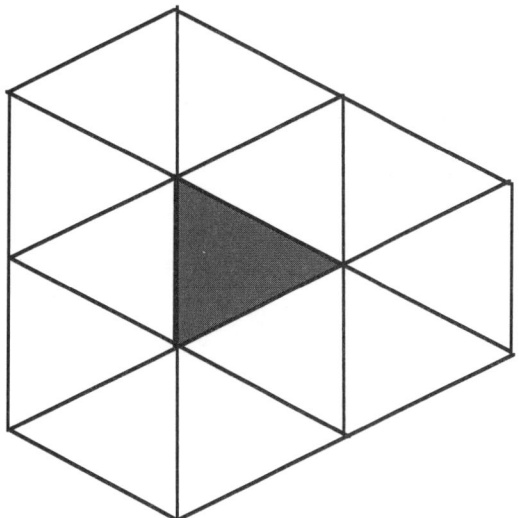
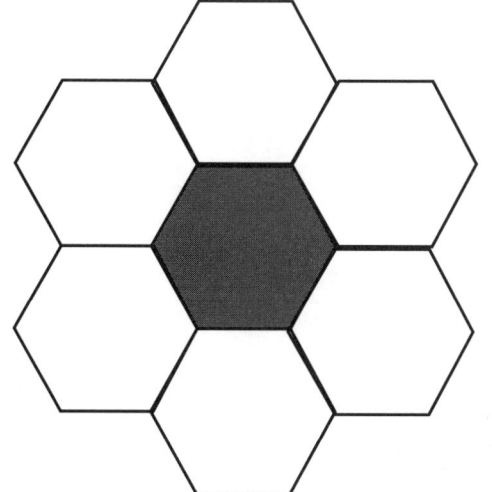
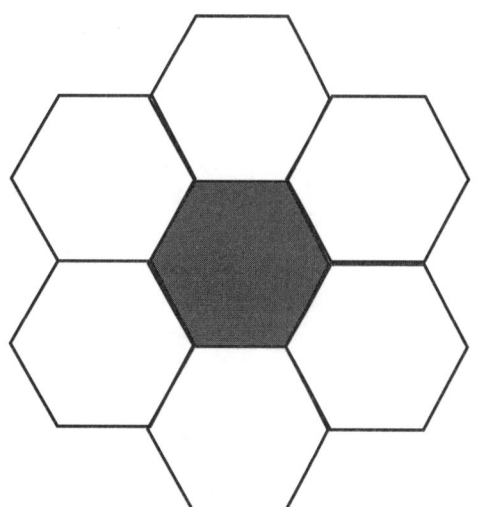

TesselMania! Math Connection

**Transparency Master
Activity 8**

Definitions of *transversal* and *diagonal*

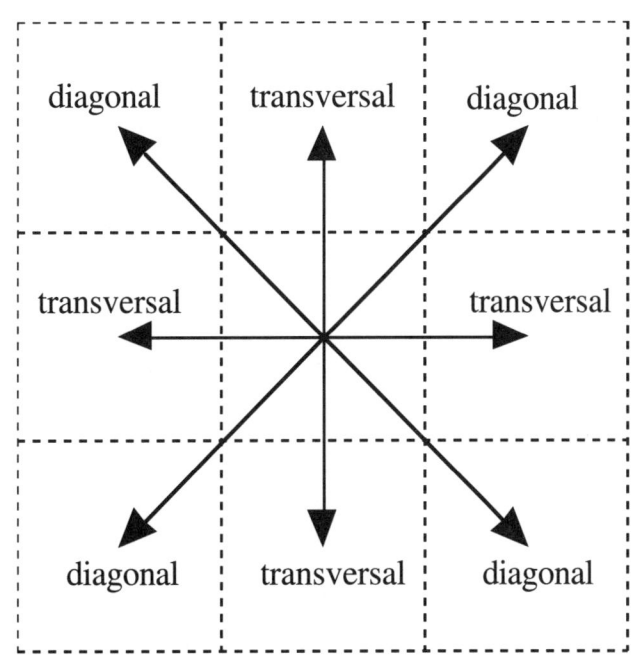

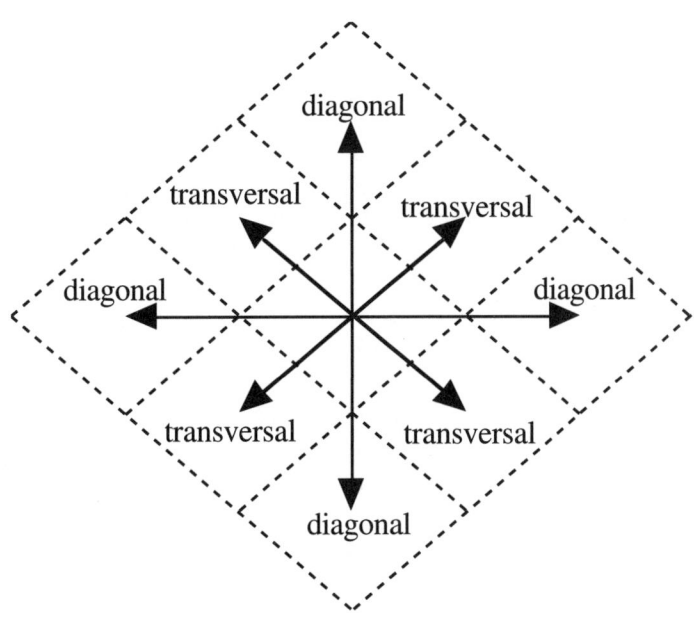

**Transparency Master
Activity 9a**

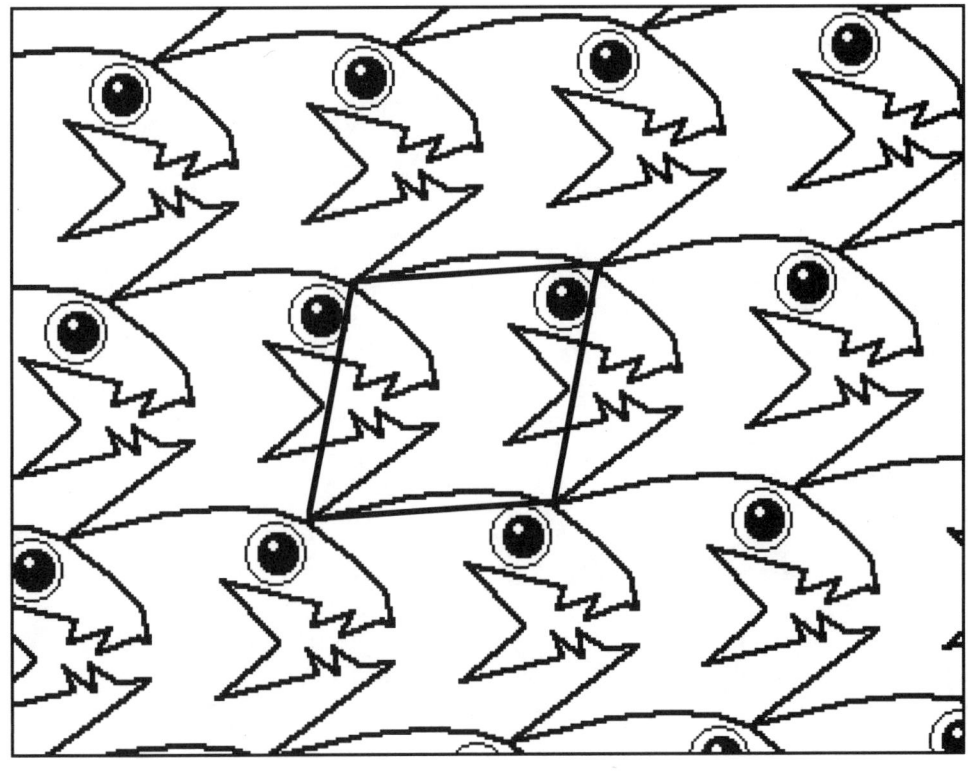

TesselMania! Math Connection ©1995 by Key Curriculum Press
Do not copy without permission.

Transparency Master
Activity 9b

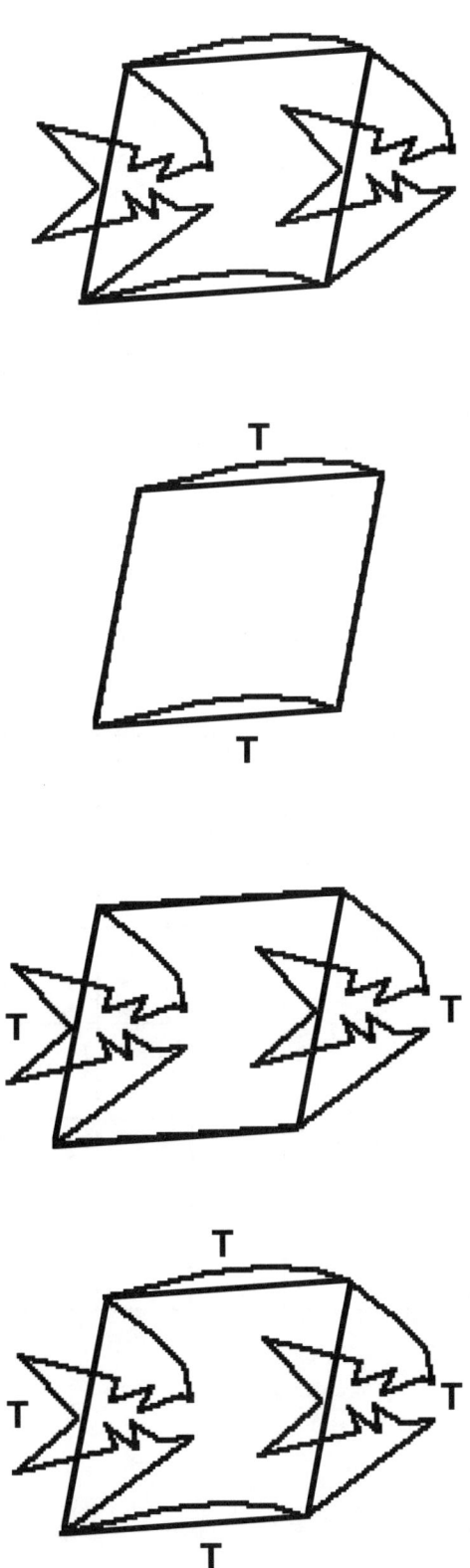

TesselMania! Math Connection

Transparency Master
Activity 11

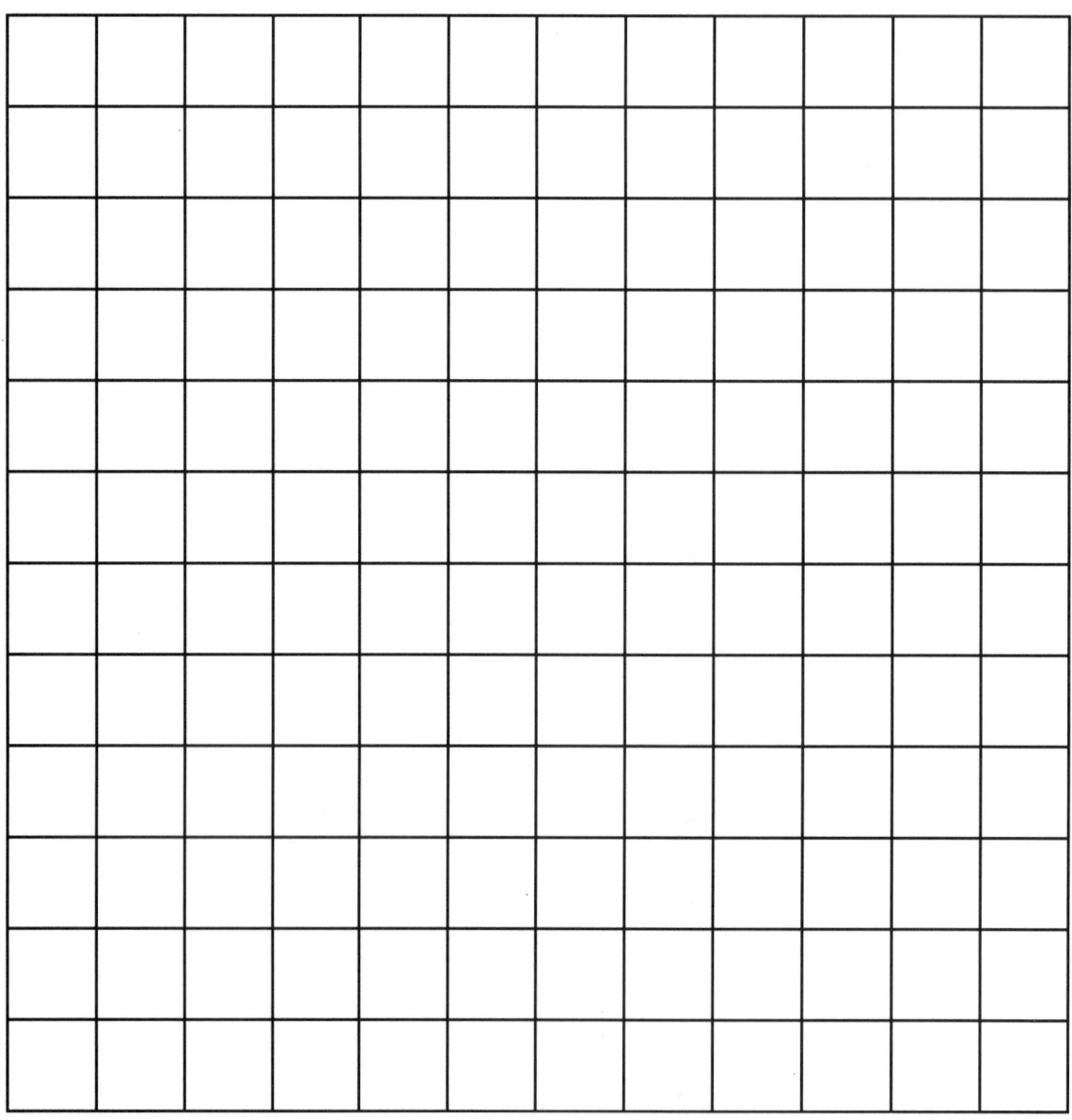

TesselMania! Math Connection